T0276901

Sisters of Prometheus

João Paulo André

Sisters of Prometheus

Unmasking Women's Achievements in Chemistry

Translated by Thomas Mindermann

 Springer

João Paulo André
Department of Chemistry
University of Minho
Braga, Portugal

ISBN 978-3-031-57135-0   ISBN 978-3-031-57136-7   (eBook)
https://doi.org/10.1007/978-3-031-57136-7

This project was translated by Thomas Mindermann.

Translation from the Portuguese language edition: "Irmãs de Prometeu-A Química no Feminino" by João Paulo André, © Gradiva publicações S. A 2022. Published by Gradiva Publicações S. A. All Rights Reserved.

This work was supported by FCT - Fundação para a Ciência e a Tecnologia

This Springer imprint is published by the registered company Springer Nature Switzerland AG
The registered company address is: Gewerbestrasse 11, 6330 Cham, Switzerland

If disposing of this product, please recycle the paper.

Preface

O fairest of creation, last and best
Of all Gods works, creature in whom excelled
Whatever can to sight or thought be formed,
Holy, divine, good, amiable, or sweet!
How art thou lost, how on a sudden lost,
Defaced, deflowered, and now to death devote?
Rather how hast thou yielded to transgress
The strict forbiddance, how to violate
The sacred fruit forbidden! Some cursed fraud
Of enemy hath beguiled thee, yet unknown

JOHN MILTON, *Paradise Lost* (1667), verses 896–905 of Book IX

In 1909, a significant milestone for gender equality was achieved when all higher education institutions in Germany opened their doors to women. However, in that same year, Nobel Prize Laureate Wilhelm Ostwald (1853–1932), a German chemist and philosopher, made a categorical statement in his book *Grosse Männer* [*Great Men*], asserting that "women of our time, regardless of race and nationality, are not suited to fundamental scientific work" [1]. Nearly 80 years later, American philosopher Sandra Harding still felt compelled to emphasize that "women have been more systematically excluded from doing serious science than from performing any other social activity except, perhaps, frontline warfare" [2]. This raises a legitimate question: why was the *official* entrance of women into the world of science delayed for so long? This question is linked with another that has recurred throughout history: Were women capable of doing science? From ancient times, the prevailing conviction had been that women's supposed weak nature rendered them unfit for rigorous reasoning.

Predating the opinions expressed by doctors and philosophers of Ancient Greece, literature had already portrayed women as irrational, malevolent, and lacking common sense. For instance, going back to the eighth century BC, Homer's *Iliad* and *Odyssey* conveyed the notion that women were perilous due to their inability to manage impulses and, for this reason, had to be kept on a short leash. Similarly, during that era, in his poem "Works and Days," Hesiod used the myth of Pandora, the first woman, to reinforce the belief that women were solely a source of problems and misfortunes [3].

In both the Hebrew Bible and the Christian New Testament, women were not only confined to the family sphere but also portrayed as being subordinate to men. God first created Adam in His image and likeness and then fashioned Eve from His rib. Furthermore, Eve was burdened with the guilt of original sin (Fig. 1). For centuries, moralists and preachers, from their pulpits,

Fig. 1 *Adam and Eve* (1526), Lucas Cranach. Courtauld Institute of Art, London

addressed women as the *weaker sex*, placing them under the weight of this biblical responsibility [4].

Saint Paul, while proclaiming, "There is neither Jew nor Greek, there is neither slave nor free; there is no man nor woman, for you are all one in Christ Jesus" (Galatians 3:28), also stated, "Let your women keep silence in the churches: for it is not permitted unto them to speak", adding that "if they will learn anything, let them ask their husbands at home" (1 Corinthians 14:34-35). In yet another epistle, he said, "I do not permit a woman to teach or to assume authority over a man" (1 Timothy 2:12). Passages such as these provided the theological and disciplinary foundations for the exclusion of women from the exercise of public functions and teaching, giving material to the Church Fathers, especially Tertullian (*c.*160–*c.*220), Jerome (*c.*347–420), and Augustine (354–430), for the perpetuation of a negative image of the female sex. Isidore of Seville (*c.*560–636) even stated that the word *mulier* (woman) was derived from *mollitia* (softness) [5]. In turn, Thomas Aquinas (1225–1274) merged the biblical concept of women as descendants of Eve, perceived as the origin of all human misfortunes, with Aristotle's belief that women were incomplete beings with the sole purpose of receiving and bearing the offspring of men. This fusion played a role in perpetuating the concept of *infirmitas mulieres* ("woman's weakness") as an unquestionable and evident reality, prompting many women to seek salvation through monastic life [6, 7].

According to the *Book of Enoch*, a set of apocryphal Old Testament texts written roughly between the third and second centuries BC, two hundred angels, led by Azazel, descended to Earth, driven by the desire for carnal pleasures. In return, they taught women the alchemical arts of metallurgy, dyeing, and the production of cosmetics and precious stones. This change brought lust, impiety, and corruption, resulting in divine wrath, and Azazel was severely punished. However, something had now become irreversible: women had acquired knowledge of alchemy [8]! Setting aside legends, the truth is that women, despite their recurrent illiteracy, played an essential role in chemical crafts, encompassing the manufacture of medicines, perfumes, cosmetics, paints, and more [9]. This aspect has garnered increasing attention from science historians in recent decades. Patricia Fara, in her work *Pandora's Breeches*, observes that today's conceptions of the evolution of science include participants whose motivations may be as trivial as improving food, health, or physical comfort for human beings, or even the mere search for wealth or recognition. This notion sharply contrasts with more traditional views that scientific development results from sporadic leaps made by isolated geniuses in a selfless search for truth [10]. Therefore, instead of focusing exclusively on great figures and their scientific discoveries, historians have also sought to investigate the work of those not belonging to universities or academies, yet contributing to the advancement of science. These were the ones who, instead of intellectual knowledge, possessed practical knowledge, meaning that they knew how to do. These were artisans such as potters, dyers, tanners, hatters, goldsmiths, etc., as well as professionals like miners, navigators, and herbalists, among others. It is under this understanding that the contribution of women to the evolution of chemistry as practitioners of the chemical crafts has created an increasing interest among historians [10, 11].

When discussing new historiographies, it is essential to mention Hélène Metzger (1889–1944), a French philosopher and historian of science who particularly focused on the history of chemistry. Her background in crystallography certainly played a role in shaping her contributions. Of Jewish origin, she fell victim to the Holocaust, dying at the age of 54 in the gas chambers of Auschwitz [12]. Thomas Kuhn, the author of *The Structure of Scientific Revolutions* (1962), a work considered a landmark in the history and philosophy of science, included Metzger in the select group of intellectuals who influenced him [13].

While keeping these considerations in mind, this book sheds light on women's historical involvement in chemical crafts, alchemy, and chemistry in general through Antiquity, the Middle Ages, the Early Modern Period, and the Age of Revolution. In Chap. 1, readers will find the first women who dedicated themselves to practices involving physicochemical processes, such as the handling and preservation of food, and the art of perfumery in the first civilizations. This chapter also delves into ancient Greek thinking and Alexandrian alchemy. Chapter 2 focuses on women's participation in European alchemy during the monastic era, addressing as well the *querelle des femmes*, a debate on women's alleged inferiority initiated by Christine de Pisan in the early fifteenth century. Dedicated to women's "books of secrets", Chap. 3 reveals the popularity of homemade medicine, cosmetics, and alchemy recipes from the Renaissance onwards. Chapter 4 explores the scientific environment in seventeenth-century England, with influences from Paracelsian iatrochemistry, mechanical philosophy, and atomism. Chapter 5 unfolds in the Age of Enlightenment, mainly in France, witnessing the dawn of modern chemistry. Lastly, Chap. 6 primarily features significant written works by or for women when formal education for them was not yet common.

Some explanations are still warranted, with the first one concerning chronology. Chapters are arranged in a sequence that, to some extent, follows chronological order. However, there might be partial overlaps in terms of their time frames with others. For instance, in Chap. 6, which focuses on chemical literature written for and by women, the narrative extends into the twentieth century. The second clarification pertains to the term "science," which is used flexibly in this book. Although its modern usage, with a strict definition, has emerged only in the nineteenth century, the term "science" has a long story, stemming from the Latin word *scientia*. It meant knowledge or understanding that was not exclusively limited to the realm of Nature and encompassed a wide range of subjects, including history, grammar, rhetoric, and even the arts.

Finally, an explanation concerning the excerpts of texts presented in the following pages (correspondence, literary works, etc.) is due. As the book was initially published in Portuguese, special care has been taken in this edition to present the excerpts of texts originally written in English in their authentic form.

Braga, Portugal   João Paulo André

References

1. Wilhelm Ostwald, *Grosse Männer*, Akademische Verlagsgesellschaft, Leipzig, 1909, p. 418
2. Sandra Harding, *The Science Question in Feminism*, Cornell University Press, Ithaca, N. Y., 1986, p. 31
3. Leigh Ann Whaley, *Women's History as Scientists—A Guide to the Debates*, ABC Clio, Santa Barbara, CA, 2003, pp. 3–4
4. M. L. King, A. Rabil Jr., "The other voice in early modern Europe", *in* Juan Luis Vives, *The Education of a Christian Woman—A Sixteenth-century Manual*, The University of Chicago Press, Chicago, 2000, pp. xiii–xiv
5. Umberto Eco (org.), *Idade Média—Bárbaros, Cristãos e Muçulmanos*, D. Quixote, Lisboa, 2010, p. 286
6. Phyllis Stock, *Better Than Rubies—A History of Women's Education*, Putnam, New York, 1978, pp. 22–23
7. Whaley, *Op. cit.* (3), pp. 21–29
8. Bruce T. Moran, *Distilling Knowledge: Alchemy, Chemistry, and the Scientific Revolution*, Harvard University Press, Cambridge, MA, 2005, p. 60
9. William Henry Hall, *The New Royal Encyclopaedia; or, Complete Modern Universal Dictionary of Arts & Sciences*, vol. 1, C. Cooke, London, 1788
10. Patricia Fara, *Pandora's Breeches—Women, Science & Power in the Enlightenment*, Pimlico, London, 2004, p. 23

11. A. Cunningham, P. Williams, "De-Centring the 'Big Picture': "The Origins of Modern Science" and the Modern Origins of Science", *The British Journal for the History of Science*, 26 (1993) 407–432
12. Marelene Rayner-Canham, Geoffrey Rayner-Canham, *Women in Chemistry— Their Changing Roles from Alchemical Times to the Mid-Twentieth Century*, Chemical Heritage Foundation, Philadelphia, 2001, pp. 194-196
13. Thomas S. Kuhn, *A Estrutura das Revoluções Científicas*, Guerra e Paz, Lisboa, 2009, p. 10

Acknowledgments

I would like to begin by expressing my gratitude to Professor Carlos Fiolhais, whose meticulous critical reading made significant contributions to the improvement of the manuscript of this book.

The idea occurred to me in Philadelphia during the last quarter of 2017 while conducting bibliographical research at both the Chemical Heritage Foundation (now the Science History Institute) and the Van Pelt Library at the University of Pennsylvania. I am particularly indebted to these institutions, with special thanks to Ronald Brashear and John Pollack.

I acknowledge Professor Margarida Casal, who was then the President of the School of Sciences at the University of Minho, for enabling my participation in the Annual Meeting of the American Chemical Society in San Francisco in May 2017. It was during this event that I had the opportunity to meet Professor Gary Patterson, who later facilitated my research at the Chemical Heritage Foundation, and to whom I am indebted.

My stay in Philadelphia was made possible by the sabbatical leave granted by the University of Minho and the leave grant from FCT (Fundação para a Ciência e a Tecnologia). I also express gratitude to FCT for the financial support provided to the Chemistry Center of the University of Minho (Ref. CQ/UM UID/QUI/00686/2019 and UID/QUI/00686/2020).

I thank Professor Ana Carneiro for patiently clarifying numerous doubts and guiding me to essential articles and books, some of which she generously provided. Appreciation also goes to Professor Raquel Gonçalves-Maia, who, upon learning of the book's structure early on, instilled in me the confidence to proceed. Likewise, I acknowledge Professor Jorge Calado, the first person I informed of my intention to pursue this project, for his unwavering support over the years.

Finally, I want to extend my profound gratitude to Gradiva, my Portuguese publisher, for not only making the Portuguese version possible but also for paving the way for this English version.

Contents

1

Perfumers and Hermetists

Devoted to her companion, who loves her in return, she grows old by his side, giving birth to beautiful and illustrious offspring. Distinguished among all women, adorned with divine grace, she has no inclination to join those who engage in discussions about Aphrodite's affairs. These are the best and most sensible women that Zeus has granted to men. However, the others, by the invention of the same god, are an eternal scourge to them.

SEMONIDES OF AMORGOS, *Types of Women*, seventh century BC

The history of women's involvement in activities related to chemical processes can be traced back to ancient times, primarily in the domains of food preparation and preservation. Women likely played roles in early pottery—and in the conversion of ores into metals. They were also responsible for preparing remedies, and with the development of the earliest civilizations, they became creators of perfumes.

In the fourth century BC, Aristotle held the belief that women were physically and intellectually inferior to men—a notion that persisted through the ages. Another lasting contribution by this Greek philosopher was his theory of the four elements, which served as the foundation for the concept of transmutation. In the early Christian era, the alchemists of Alexandria embarked on extensive exploration of this concept, ultimately seeking to transform base metals into gold.

The heroes of this chapter include the prehistoric women, as well as figures such as Tapputi, a perfumer from Babylon, and Maria the Jewess, an alchemist likely from Alexandria.

© The Author(s), under exclusive license to Springer Nature Switzerland AG 2024
J. P. André, *Sisters of Prometheus*, https://doi.org/10.1007/978-3-031-57136-7_1

1.1   Origins

Before the advent of hunting, humans were primarily engaged in food gathering as their main subsistence activity. Women played a crucial role in performing tasks related to this endeavor, leading them to design tools and methods for collecting, preparing, and preserving food. Among the earliest instruments were sticks, levers, and hand axes, as well as simple stones for extracting roots, scraping, and grinding plant products. Over time, these primitive tools evolved into more advanced equipment, including mortars and pestles, along with rudimentary systems for milling grains and seeds. As hunting activities became more prominent, women acquired skills in carving meat, processing animal products, tanning hides, and using leather for various purposes. These skills were likely followed by the invention of the needle and the discovery of natural pigments (not necessarily in this order).

It is highly plausible that our female ancestors may have been involved in the discovery and development of pottery. If this were the case, transitioning from using their kilns for firing clay to using them for extracting metals from ores might have been a small step. Additionally, they assumed roles as midwives, healers, and surgeons, applying their knowledge of the medicinal properties of plants acquired through food collection. It is not far from the truth to say that therapeutic practices made limited progress from the time when these prehistoric women relied on herbs and roots until the discovery of sulfonamides and penicillin in the twentieth century [1].

From the third millennium BC onward, the civilizations inhabiting the region extending from the Nile to the Euphrates raised metalworking techniques to truly admirable levels. They were also skilled in glazing ceramic pieces and in the production of glass, a material that Egyptians began producing on a large scale in the fourteenth century BC. By then, the process of obtaining indigo blue from plants of the *Indigofera* genus and the use of mordants [2] in dyeing fabrics were also common in Egypt. The production of beer, medicines, and perfumes was equally important in the land of the pharaohs, with the particularity of usually being done by women (Fig. 1.1) [3]. Perfumery reached a high level of development in Babylon, the great Mesopotamian city founded in 2300 BC on the banks of the Euphrates. It was in Babylon that some of the classic techniques for extracting essential oils emerged, namely, pressing and maceration. The essential oils were necessary for scented waters and lotions, as well as ointments and other preparations for medicinal, magical, and religious purposes [4].

Fig. 1.1 Preparation of lily perfume by women; fragment from a Fourth Dynasty Egyptian tomb; *c.* 2700–2200 B.C. Louvre Museum, Paris

The cuneiform writing found on clay tablets dating back to the thirteenth century BC has yielded valuable insights into the techniques employed in Babylonian perfumery. These ancient texts not only disclose the solvents used but also outline the necessary equipment for this craft. This equipment included an array of items such as pots in various shapes and sizes made from clay, glass, or metal, measuring cups, basins, sieves, flasks, furnaces, and possibly sublimation devices. Among the technical information gleaned from these tablets, we also find the name of one of the perfumers, Tapputi-Belatekallim, where "Belatekallim" indicates her role as a supervisor in the royal palace. Furthermore, from another woman, likely an assistant to Tapputi, we learned the latter part of her name, "ninu." In her perfumery, Tapputi used not only myrrh and balms but also botanical species from the *Cyperus* genus and *Acorus calamus* [5–7].

A significant part of Western civilization came from Egypt and Mesopotamia, but it was the ancient Greeks who endowed it with an intellectual splendor that still surprises us today. They were the true creators of philosophy, science, and mathematics (arithmetic and geometry already existed in Egypt and Babylon but only with practical rules). Based on rationality and logic, Greek philosophers crafted theories to explain the origins of the universe, the composition of material bodies, and their transformations. According to the pre-Socratic philosophers, everything had its origin in a primordial matter. Thales of Miletus (*c.*625–*c.*545 BC) believed this matter to be water, while his disciple Anaximander (*c.*610–*c.*546 BC), also from Miletus, posited that the material basis of all things was something formless and indeterminate, which he called *apeiron*. For Anaximenes, also from Miletus (*c.*585–*c.*528 BC), the origin of everything was air. In the view of Heraclitus of Ephesus (*c.*540–*c.*480 BC), it was fire, and for Xenophanes of Colophon (*c.*570–*c.*475 BC), it was earth. In an attempt at unification, Empedocles of Agrigento (*c.*490–*c.*435 BC) considered water, air, fire, and earth to be the "roots of all things."

Shortly thereafter, Leucippus and his disciple Democritus of Abdera (*c.*460–*c.*370 BC) argued that everything was made up of atoms. In Greek, the word *atomos* meant "indivisible." These were corpuscular entities that, in an infinite number and variety and in constant vortices, filled the void, which was a prerequisite for movement. In their permanent whirlwind—beyond any external cause, as it was the ultimate cause of everything—atoms could unite among themselves, giving rise to all forms of matter. The properties of matter would be determined by the sizes and shapes of their constituting atoms.

Epicurus of Samos (341–270 BC), the founder of the Epicurean School, shared an atomist view of the world. One of his followers, the Roman poet and philosopher Titus Lucretius Caro (*c.*99–*c.*55 BC), made significant contributions to the dissemination of atomism in the Latin-speaking world with his remarkable poem *De rerum natura* [*On the Nature of Things*]. The Epicureans, who believed that the soul was also composed of atoms, introduced new dimensions to atomism. They proposed that atoms did not move in vortices but followed parallel rectilinear trajectories, similar to bodies in free fall. Occasionally, they could undergo a small and unpredictable swerve, known as the *clinamen*, leading to collisions with other atoms, allowing for the formation of matter. Atomism was a doctrine that later also acquired a moral and ethical dimension, particularly linking human free will—liberated from the constraints imposed by religions, superstitions, or ignorance—to the concept of the *clinamen*. Although it eventually faded into obscurity, atomism experienced a revival starting in the fifteenth century, as will be explored in Chap. 4.

In the fifth century BC, the Athenian Socrates (*c.*470–399 BC) left an indelible mark on philosophy. He advocated deductive reasoning, always commencing with an irrefutable premise, and held disdain for experimentation, believing that reflection alone was sufficient to comprehend the world. His interests, however, focused more on the nature of the human being than that of the universe, advising individuals to "know thyself." Socrates considered natural philosophy too speculative (a tendency from which mathematics was exempt), but the significance he placed on aspects such as the need for clarity in definitions and classifications, logical argumentation, respect for order, and rational skepticism had a lasting impact on the future of science. Despite being sentenced to death for allegedly corrupting the ideas of Athenian youth, his influence continued to flourish in fertile ground, particularly through another Athenian, Plato (c.427–c.347 BC), his most brilliant disciple.

Considering that atheism was incompatible with understanding the universe and that natural laws were believed to be subject to divine principles, Plato's natural philosophy marked a departure from that of his predecessors. Nevertheless, he, like them, argued that everything was formed by the four material principles of Empedocles, combined in varying proportions. He was also the first to refer to them as "elements" (*stoicheia*).

In his theory of ideas (or forms), which posited that an observed object was merely an imperfect copy or reflection of an idea, with the idea being the only true reality, Plato argued that the senses were not reliable. For this reason, he contended that knowledge could only be achieved through deductive reasoning and not through direct observation, which explains the little importance he placed on experimentation. Influenced by the mathematical spirit of the Pythagorean School, founded by Pythagoras in the sixth century BC and which held that the cosmos could be explained through arithmetic and geometric relationships, Plato associated ideas with numbers.

In *Timaeus* (c.360 BC), a dialogue that can be considered one of the earliest treatises on chemistry, as it includes a discussion on the composition of organic and inorganic bodies, Plato expounded his vision of cosmogony. He associated each one of the primordial elements with a regular geometric solid: fire to the tetrahedron, air to the octahedron, water to the icosahedron, and the earth to the cube. He further suggested the existence of a fifth element, the ether or quintessence, present in celestial space, which he associated with the dodecahedron. These polyhedral, known as Platonic solids, were believed by Plato to be convertible elements. According to him, it was possible to convert elements into each other by resolving the respective polyhedra into right-angled triangles and reassociating them. However, due to the impossibility of solving the pentagonal faces of the dodecahedron, this polyhedron was considered a separate case with a divine connotation.

The concept of the soul was of central importance in Plato's thought. Recognizing that the soul was devoid of gender, he advocated social equity for women and promoted similar education for both boys and girls, although this was typically restricted to those from the ruling class. For this reason, despite his sexism, he is sometimes considered one of the first Western intellectuals who championed the female gender [8].

Plato, who founded the School of Athens around 387 BC, paved the way for his disciple Aristotle (384–322 BC). In turn, Aristotle established the famous Lyceum in 335 BC, where the Peripatetic School flourished. Although he followed in the footsteps of his master, Aristotle was more favorable to the observation of natural phenomena, as he did not conceive that only ideas could form reality. Nevertheless, in his systematic approach to understanding the world, Aristotle prioritized the use of syllogism, a form of deductive reasoning based on a priori premises considered to be true, leading to deduced conclusions. Consequently, there was no room for experimentation in his methodology. Aristotle's physics (*physis* was the Greek term for the final cause of something, or its purpose, although it is often translated as "nature") is essentially a metaphysical exploration of the nature of things rather than an inquiry into their laws. By rejecting the notion that everything was composed of atoms and empty space, Aristotle supported the concept of a fundamental substance called *hyle*. Similar to clay shaped by a potter, *hyle* possessed the capacity to assume diverse forms. However, these forms would develop from within, akin to an organic growth. In his treatise *On Generation and Corruption*, he argued that it was from the union of form with pairs of sensible qualities (hot, cold, dry, and wet) that the four elements of cosmogony resulted, as illustrated in the diagram in Fig. 1.2. All bodies were generated from the elements combined in varying proportions, except for celestial bodies, which, considered perfect, contained the ether or quintessence in their constitution. Aristotle observed that any two adjacent elements in the diagram shared a common sensible quality, leading him to view them as interconvertible, thus perpetuating the concept of transubstantiation introduced by his master.

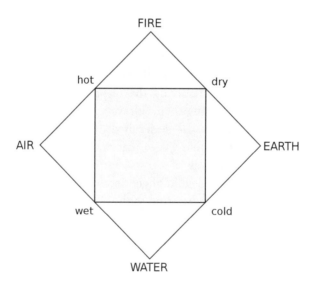

Fig. 1.2 The four Aristotelian elements and their shared properties

Hellenistic medicine also reflected Aristotelian doctrine. In the *Corpus Hippocraticum*, a collection of writings attributed to Hippocrates (*c.*460–*c.*370 BC) and his followers, the notion of assigning a natural cause to each specific illness emerged for the first time, thereby separating medical practice from religion and magic. Diseases were considered as the body's natural reactions rather than punishments from the gods. The *Corpus Hippocraticum* expounds the theory of the four humors (fluids) of the body—blood, yellow bile, phlegm, and black bile—originating, respectively, from the heart, liver, respiratory system, and spleen. These humors were associated with the fundamental qualities of the four Aristotelian elements: blood, hot and moist like air; phlegm, cold and wet as water; yellow bile, hot and dry as fire; and black bile, cold and dry as the earth. According to Hippocrates, often referred to as the "father of Medicine," a healthy body relied on the close balance between the four humors. Any excess or shortage of these humors resulted in specific symptoms and disorders. Furthermore, the humoral theory suggested that the predominance of a particular humor in a person's constitution was also responsible for their temperamental characteristics. Galen of Pergamum (129–*c.*216 AD), the physician responsible for the perpetuation of this theory for many centuries, classified four fundamental temperaments (Fig. 1.3)—phlegmatic (associated with phlegm), sanguine (blood), choleric (yellow bile), and melancholic (black bile). As a result, men and women were believed to have different temperaments. Men, being hot and dry, were considered choleric, which accounted for their strength and intelligence, while women, being cold and wet, were labeled phlegmatic and thus considered weak and irrational.

Aristotelian doctrine made its way into Jewish, Christian, and Islamic thought. Some authors attribute the enduring influence of his ideas on European culture for almost two thousand years to the fact that he was the tutor of Alexander the Great (356–323 BC). Aristotle's views on the role and status of women in the society were shaped by biological considerations, leaving a lasting impact. He held the belief that women were inferior to men—physically, intellectually, and socially. In his work *Rhetoric*, he asserted the importance of women's happiness being identical to that of men, as a society could not be happy if they were not. However, in *Politics*, he regarded women as subaltern and incomplete beings. Even in matters of reproduction, Aristotle underestimated the role of women, considering their reproductive contribution to be inferior to that of men due to a lower degree of heating and cooking of their blood, which he attributed to their colder nature. In his view, women merely provided the raw material for the development of new beings, while it was men's semen that possessed the principle of movement necessary for generating the form [8–10].

Fig. 1.3 The four temperaments and their humors. Illustration from Leonhard Thurneysser's *Quinta Essentia* (1570)

In the Hellenic world, women were barred from participating in any public events. While boys had access to knowledge in the gymnasium, girls were confined to the gynaeceum and, at best, could only learn to read and write. Educating a woman, as Menander (fourth and third centuries BC), the author of several Greek comedies, ironically put it, was like "giving more poison to a snake" [11]. On the island of Lesbos, the poet Sappho (*c.*630–*c.*570 BC) ran a school for young women where they could learn philosophy, poetry, music, and dance. However, except for rare exceptions like this one, or the cases of Aspasia of Miletus (*c.*470–400 BC) and Arete of Cyrene (fifth and fourth centuries BC) in philosophy, or Aglaonice of Thessaly (second century BC) in astronomy, women in ancient Greece were excluded from all scientific and philosophical discussions. It is worth noting that Aglaonice was often regarded as a sorceress due to her ability to make the moon disappear. However, her true skill lay in predicting eclipses [12, 13].

1.2   Alexandria

In spite of the importance attributed to the family, which contributed to some elevation of the status of women, the Romans perpetuated the conviction of female inferiority. Unlike Greek girls, upper-class Roman girls were instructed, primarily to prepare them for their roles as mothers and to equip them to educate male children for civic life. Many of them shared their brothers' tutors. There were also schools for girls from less affluent families where they could receive a classical literary education. After the fall of the Western Roman Empire in 476 AD, joint education for boys and girls would only reemerge in the Italian courts of the Renaissance.

While women in Rome enjoyed more freedom than their counterparts in Greece, none of them became prominent in scientific fields. In fact, Roman civilization, as a whole, did not value science; instead, it prioritized technology. Nevertheless, during the late Roman Empire, several women distinguished themselves in the field of alchemy. Interestingly, this pursuit did not take place on the Italian Peninsula but in Egypt, which had been a Roman province since 30 BC. In particular, it occurred in Alexandria, the bustling metropolis situated in the Nile Delta.

Founded by Alexander the Great in 331 BC, Alexandria eventually evolved into a hub of Hellenistic culture, surpassing even Athens. During the first centuries of the Christian era, women found there favorable conditions to dedicate themselves to liberal activities, including those that contributed to the genesis of alchemy. The vast city presented a unique blend of cultural and religious eclecticism, amalgamating ancient Egyptian heritage with Greco-Roman culture. Moreover, it became a melting pot where Christianity and Judaism intersected with diverse, often esoteric, cults. Thus, Alexandria witnessed the merging of two significant sources of knowledge: one rooted in the traditions of metallurgy, textile dyeing, and glassmaking and the other in Greek philosophical thought, which had also been open to influences from Asian doctrines and mystical currents.

As Platonism transitioned into Neoplatonism, a doctrine championed by figures like Plotinus (c.204–270) who posited that all reality stemmed from a single principle, known as the One, Greek philosophy underwent a transformation. It became infused with elements of mysticism and occultism due to the proliferation of philosophical-religious sects. In this context, the association of the practices of imitating noble metals and precious stones (described in the Leiden and Stockholm Papyruses, c.300) to doctrines on the nature of matter might have resulted in the desire to obtain gold from base metals.

This became known as "the Work," "the divine Art," or "the sacred Art." Alchemy was possibly born in this cultural crucible. Etymologically, the term results from the juxtaposition of the Arabic prefix *al* with *chem*, a word that most probably comes from the Greek *chemeia*, meaning the "art of melting" metals. There is also the more traditional interpretation that the latter derives from the Coptic word *kheme*, meaning "blackness," alluding to an old designation of the nation of the pharaohs as "the land of the black soil," due to the silt of the Nile. In that sense, alchemy would be referred to as "the Egyptian art" [14].

One of the foundational concepts in Greek-Egyptian alchemical thought was the idea of the unity of the whole, symbolized by Ouroboros, a serpent coiled in a circle, biting its own tail (Fig. 1.4), representing the cosmos. At the time, the prevailing Aristotelian idea was that all things originated from a single primordial matter, composed of the four elements combined in varying proportions and in a constant state of change. This belief supported the notion of transmuting substances into one another.

Zosimos of Panopolis is considered one of the foremost exponents of Greco-Egyptian alchemy. He wrote a significant number of texts between the end of the third century and the beginning of the fourth century, but much of his work has been lost, surviving only in fragmentary form, either in the original Greek language or translated into Syriac, Arabic, and Latin. In the late nineteenth century, the French chemist Marcellin Berthelot translated some of these texts [15]. Zosimos' writings are scattered throughout the *Corpus Alchemicum Graecum*, a miscellany of Greek alchemical texts assembled by Byzantine scholars between the seventh and eleventh centuries. They were propagated by numerous medieval manuscripts [16–19], which contain drawings and descriptions of experimental setups and procedures, alongside mystical and psychological elements with a gnostic-hermetic inclination. Zosimos goes so far as to suggest that angels descended to Earth, drawn by the beauty of women, to instruct humanity about the workings of Nature, making a clear allusion to the Book of Enoch (as discussed in the Preface). Furthermore, references to the female sex are not lacking in his work. Some of his texts, sometimes referred to as *Cheirokmeta* or "things made by hand," take the form of letters addressed to his sister, Theosebia, who was likely an alchemist herself. Interestingly, in certain excerpts, Zosimos warns her against forming relationships with individuals he considers uneducated, fearing their negative influence on her work. Among them was a woman known as Pafuntia the Virgin, a priestess who likely belonged to a distinct school or sect [20].

Fig. 1.4 Ouroboros. Illustration from *Chrysopoeia*, a manuscript from the tenth to eleventh century assigned to Cleopatra, an alchemist from the third or fourth century. The inscription reads: "The whole is one," expressing the unity of all things

From the available sources, it is difficult to assess the originality of Zosimos, as he evidently based his work on the contributions of alchemists who preceded him. Notably, he frequently mentions and holds in special reverence figures such as Pseudo-Democritus, Hermes Trismegistus, and Maria the Jewess, among others. Maria, also known as Maria the Hebrew or the Prophetess, was a practitioner of the Hermetic art, and she is believed to have lived in Alexandria during the second or third century AD. A century or two later, she was already considered the founding mother of alchemy and one of its most prominent figures, as attested by Zosimos's writings.

A significant part of Maria's teachings can be found in a treatise on Arabic alchemy by al-Habib (unknown date), which recounts the dialogue she allegedly had with a figure named Aros, also identified as "King Horus." In a later Latin version, Maria, while teaching Aros how to make gold, says, "Take the white, clear, precious herb which grows upon the small mountains, and pound it freshly as it is in its hour; and it is the body which truly does not become fugacious with fire." Aros asks, "Is that not the Stone of Truth?." And Maria replies, "It is." A few lines later, she returns to the subject and adds some details: "Take that clear body which grows on the small mountains, which is not subject to either putrefaction or movement, and pound it with gum elsaron and with the two fumes […].” Maria further explains to Aros that “the two aforementioned fumes are the roots of this art.” The enumeration of the basic ingredients of the Great Art—the two fumes (white and red, representing, respectively, mercury and sulfur) and the herb that grows in the small mountains—is recurrent in alchemical literature. Michael Maier, an important figure in early seventeenth-century alchemy, includes in his *Symbola Aureae Mensae Duodecim Nationum* a section entitled “Mariae Hebraeae Symbolum” in which he places Maria at the same level as Hermes Trismegistus. This work contains a representation of her (Fig. 1.5): an imposing woman in a cloak and veil pointing with her left hand to a small mound where the “white grass” grows. At the bottom, there is an urn from which two columns of fumes emanate and separate to surround the plant, joining two other fumes that descend from an inverted urn suspended in the air. The “white herb” could be *Botrychium lunaria*, a plant of special significance in alchemy, believed to come from the Moon and to be impregnated with a “celestial vitality.” Possessing all kinds of healing and magical qualities, it represented the phase of white in alchemy, which is the very first color transformation from the base black, followed by transformation into yellow and then red [21].

Fig. 1.5 Mary the Jewess. Engraving from Michael Maier's alchemical treatise *Symbola Aureae Mensae Duodecim Nationum* (1617). Courtesy of Adam McLean

In the doctrine of Maria the Jewess, two main ideas stand out: first, the belief that all bodies and substances found in Nature are essentially one; and second, the analogy between human beings and metals. According to her, metals, like humans, had body, soul, and spirit, and they were classified as either male or female. The hidden inner nature of metals could be revealed through a complex alchemical process that would ultimately lead to their transformation into gold, known as *iosis*. Maria claimed that this great mystery was revealed to her by the grace of God and should only be known by the Jews. As it appears in various Greek manuscripts, she recommended that Gentile peoples should not touch the philosopher's stone with their hands, stating, "You are not of our race, you are not of the race of Abraham" [21].

Maria the Jewess is credited with inventing several alchemical furnaces and experimental devices that remained in use until the eighteenth century, virtually unchanged. Among these apparatuses was the double boiler, known in Latin as *balneum Mariae* (*bain-marie* in French, *banho-maria* in Portuguese, *bagnomaria* in Italian), which is still used today in both kitchen and laboratory settings. Other notable inventions attributed to her include the *tribikos* (a tri-tubular still), the *ambix* for distillations, and the *kerotakis* (Fig. 1.6), likely intended for the vaporization and sublimation of substances, as well as reactions of their vapors with metals [22]. The *kerotakis* might have derived from the metallic palette used by ancient Greek painters to mix pigments with wax (*keros*), which had to be kept warm to prevent solidification. Over time, it evolved into an apparatus placed on a furnace and consisting of three parts, as represented in the figure: a plate for the vaporizable substances, a palette for the metals to be converted into gold, and a convex cover to condense the

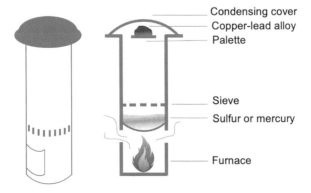

Condensing cover
Copper-lead alloy
Palette

Sieve
Sulfur or mercury

Furnace

Fig. 1.6 *The kerotakis* (modern representation)

formed vapors. Maria seems to have emphasized the use of alloys of copper, iron, zinc, and lead ("four in one"), referred to as "our lead." However, it could also include antimony or some metal sulfides [21, 23].

Interestingly, the perfect seal obtained in the *kerotakis*, in association with Hermes Trismegistus, led to the expression "hermetically closed." Hermes Trismegistus, whose name means "three times great," was originally a figure created as a symbiosis between Thoth (the Egyptian god of wisdom) and Hermes (the Greek deity of communication). Later, he came to be considered the wise founder of the Hermetic disciplines, namely, alchemy. Between the Renaissance and the beginning of the seventeenth century, it was believed that he had been a contemporary of Moses and the oldest prophet of God, as well as the author of the *Corpus Hermeticum*, the writings that form the basis of Greco-Egyptian Hermeticism. However, it is now known that these texts, which express an eclectic mix of Egyptian, Greek, Roman, and Christian thought with elements of Platonism, Neoplatonism, Gnosticism, and Stoicism, were written anonymously in the first centuries of our era [24].

In the Biblioteca Marciana in Venice, one can find manuscripts dating back to the tenth or eleventh century that consist of transcriptions in Greek of much older texts, including alchemical treatises by several authors. Among these works are those by Zosimos and those from a woman named Cleopatra, as well. The latter, who should not be confused with the famous Egyptian monarch (Cleopatra was a common name in Ancient Egypt), may have also been an alchemist in Alexandria during the early Christian era and is believed to have been associated with the school of Maria the Jewess. In her treatise on *chrysopoeia* (the transmutation of base metals into gold), there is a page with drawings of experimental apparatus used in alchemical procedures, along with a depiction of Ouroboros [23].

Maria the Jewess and Cleopatra are said to have been among the first and last experimental alchemists of their time. In 296, under the edict of the Roman Emperor Diocletian, alchemy practitioners in Alexandria faced persecution, and their treatises were destroyed [25, 26]. In the following century, Hypatia (*c.*355–415), a prominent Neoplatonic mathematician and astronomer in the city of the Nile Delta, is credited with inventing a hydrometer for liquids, as described in a letter by Synesius of Cyrene (370–413). Hypatia (Fig. 1.7), the first woman of science whose biography has been well-documented, tragically became a victim of a brutal murder carried out by a group of Christian monks [27]. Her death coincided with the decline of the Roman Empire. Subsequently, Alexandria came under Islamic rule in 640, and both the Byzantine Empire and Islam served as important repositories of Ancient Greek knowledge. The Arabs, in particular, played a crucial role in

Fig. 1.7 Hypatia of Alexandria. Fictional portrait by Jules Maurice Gaspard, 1908

preserving the thinking of classical authors through translations, analysis, and annotations. This formed the foundation of their scientific and philosophical thinking, which also integrated ideas from China and India [28–31].

1.3   In the East

Despite the fact that the ancient Chinese and Indian civilizations shared a cosmogony based on primordial elements, similar to the Greeks, and believed in the transmutation of base metals into gold, the alchemical systems they developed were quite different. In China, according to Taoist philosophical and religious doctrine, substances were considered to be formed by five elements—metal, wood, earth, water, and fire—and everything in the universe obeyed the yin-yang dualism, meaning two opposing fundamental forces. For instance, all metals and minerals were believed to be essentially the same thing, differing only in the relative amount of yin and yang (the first being the feminine principle associated with the moon, darkness, and passivity; the second being masculine, connoted with the sun, light, and activity). Another characteristic of Taoism was the rejection of the pursuit of knowledge, and humans should limit themselves to a basic and peaceful existence, dedicated to the contemplation of Nature. The Taoists believed that this was the only way to achieve a state of inner peace that prolonged life. However, over time, these ideals shifted as the Taoist pursuit of longevity became associated with the quest for immortality, which was believed to be achieved not through contemplative attitude but through an elixir. The secret was its content in gold, an incorruptible and eternal metal that, once in the body, led to a state of perpetual life. Thus, it was up to the alchemists to make gold drinkable. Often, the elixir, or immortality pill, contained mercury, which was believed to transmute into gold [32, 33].

The first Chinese treatise on alchemy dates back to the year 140 of our era and was written by Wei Boyang. In this work, the author defends the ingestion of gold for achieving eternal life and describes potable preparations of this metal. However, the most significant Chinese alchemist emerged in the fourth century: Ge Hong (*c.* 280–340). As a Taoist philosopher and believer in magic, he authored *Baopuzi*, a detailed treatise containing several recipes for producing gold from metals like mercury and lead. This work also mentions various female figures practicing alchemy. His own wife, Pao Ku, is said to have dedicated herself to this activity.

During ancient times in China, the society was predominantly matriarchal, and women were highly regarded. However, around 1100 BC, the system shifted to a patriarchal structure, leading to the degradation of women's status. Moreover, the teachings of Confucius, present in the ethical basis of traditional Chinese society, did not favor women's position. They were subjected to parental authority before being handed over to their husbands, with little authority beyond their role as mothers or within the family. Despite the repression of female intellectual activity, exceptions did exist in areas like arts, letters, politics, and even military affairs. In the field of science, there were also a few exceptions, although not many.

The Chinese women of antiquity who engaged in scientific pursuits and whose names have come down to us mostly came from cultured families and received a thorough education. One such figure was Fang, an alchemist who lived around the first century BC. Hailing from a family of alchemy scholars, she learned the "art of yellow and white" (a term denoting alchemy) from one of Emperor Wu of Han's favorite wives. Fang is credited with the ability to convert mercury into silver, which may have been the extraction of silver from metal alloys due to its tendency to amalgamate with mercury. Unfortunately, her life took a tragic turn when her husband, Cheng Wei, who was also interested in alchemy, forced her to reveal the secret of her feat, threatening to dismember her. Despite his cruel intimidations, she refused to disclose the information and fell into dementia. One day, in a state of despair, she ran out of the house, totally naked, and committed suicide by drowning in a lake of mud. Upon learning of these events, the emperor had Cheng Wei executed. This is believed to have occurred around 45 BC, as reported by Ge Hong [34–36].

In India, the *Vedas*—a collection of sacred Hindu texts from *c.*1000 BC, whose Sanskrit title means "knowledge"—admitted the existence of five primordial elements: earth, water, air, ether, and light. The Vedic doctrines also defended atomism, originally proposed in Indian culture by Kanada, a philosopher who lived between the sixth and second centuries BC. The Indian concept of atom (*anu*, meaning "small") was thought as something spherical and indestructible, possessing properties such as color, taste, and smell, and associating itself first in pairs and then in aggregates of pairs.

A striking aspect of the *Vedas* is the association of gold with longevity. The idea of metals being transmuted into gold was also present in Indian alchemy, however, obtaining elixirs made of gold was considered much more important than becoming rich. A fundamental difference from Chinese alchemy was that such preparations with gold were used for therapeutic purposes and not to achieve immortality—for that purpose, there was Tantrism, a philosophical system with aspects common to Taoism. The *rasa shastra*, which in Ayurvedic medicine (traditional Indian medicine) is the "science of mercury," inherited this alchemical tradition. The habit of decorating food with gold leaf, particularly at banquets, is still common in India today [37, 38].

Although specific information about Indian alchemists before the seventh century is missing, the considerable number of Hindu women who distinguished themselves in fields such as politics, justice, accounting, and even the military suggests that alchemical practices may not have been unfamiliar to them, similar to what was observed in China and Alexandria [39].

References and Notes

1. Margaret Alic, *Hypathia's Heritage—A History of Women in Science from Antiquity to the Late Nineteenth Century*, The Women's Press, London, 1990, pp. 12–13
2. Metal salts that bind the dye to the fibers, giving them different colors depending on the nature of the metal.
3. P. Loyson, "Chemistry in the time of the pharaohs", *J. Chem. Educ.*, 88 (2011) 146–150

4. Charles S. Sell (Ed.), *The Chemistry of Fragances—From Perfumer to Consumer*, RSC Publishing, Cambridge, 2006, pp. 4–5

5. Martin Levey, *Chemistry and Chemical Technology in Ancient Mesopotamia*, Elsevier, Amsterdam, 1959, pp. 142–144

6. M. Levey, "Perfumery in Ancient Babylon", *J. Chem. Educ.*, 31 (1954) 373–375

7. M. Levey, "Some Chemical Apparatus of Ancient Mesopotamia", *J. Chem. Educ.*, 32 (1955) 180–183

8. Leigh Ann Whaley, *Women's History as Scientists—A Guide to the Debates*, ABC Clio, Santa Barbara, CA, 2003, pp. 8–17

9. M. G. F. Schalcher, "Consideraçós sobre o tema da mulher no pensamento de Aristóteles", *Phoînix*, 4 (1998) 331–344)

10. M. Lopes, "Para a história conceitual da discriminação da mulher", *Cadernos de Filosofia Alemã*, 15 (2010) 81–96

11. G. K. Clabaugh, "A History of Male Attitudes toward Educating Women", *Educational Horizons*, 88 (2010) 164–178

12. C. L. Herzenberg, S. V. Meschel, J. A. Altena, "Women Scientists and Physicians of Antiquity and the Middle Ages", *J. Chem. Educ.*, 68 (1991) 101–105

13. Except if indicated, the section devoted to Greek philosophical thought was based on the following books: António M. Amorim da Costa, *Introdução à História e Filosofia da Ciências*, Europa América, Lisboa, 2002, pp. 30–65; Bertrand Russel, *História da Filosofia Ocidental*, vol. 1, Círculo de Leitores, Lisboa, 1978, pp. 20–165; pp. 191–198; W. K. C. Guthrie, *The Greek Philosophers—From Tales to Aristotle*, Routledge, Abingdon, 2013.

14. "The Etymology of "Chemistry"", *J. Chem. Educ.*, 7 (1930) 652

15. Marcellin Berthelot, *Collection des Anciens Alchimistes Grecs*, G. Steinheil, Paris,1888, 3 vols.

16. H. S. Elkhadem, "A Translation of a Zosimos Text in an Arabic Alchemy Book", *Journal of the Washington Academy of Sciences* 84 (1996) 168–178

17. M. Mertens, "Project for a new edition of Zosimus Panopolis", *in Alchemy Revisited: Proceedings of the International Conference on the History of Alchemy at the University of Groningen*, Z.R. W. M. von Martels (ed.), E. J. Brill, Leiden, 1990

18. M. Mertens, "Zosimus, alchemist", *in* R. S. Bagnall, K. Brodersen, C. B. Champion, A. Erskine, S. R. Huebner (eds.), *The Encyclopedia of Ancient History*, First Edition, John Wiley & Sons, Hoboken, 2013, pp. 7185–7186

19. "Zosimos of Panopolis"; https://www.encyclopedia.com/science/dictionaries-thesauruses-pictures-and-press-releases/zosimos-panopolis

20. Marilyn Ogilvie, Joy Harvey e Margaret Rossiter (eds), *The Biographical Dictionary of Women in Science: Pioneering Lives From Ancient Times to the Mid-20th Century*, Routledge, Oxford, 2000, p. 978

21. R. Patai, "Maria the Jewess-founding mother of alchemy", *Ambix*, 29 (1982) 177–197

22. J. R. Partington, "The *kerotakis* apparatus", *Nature*, 159 (1947) 784

23. F. Sherwood Taylor, "A Survey of Greek Alchemy", *J. Hell. Stud.*, 50 (1930) 109–139
24. Stanton J. Linden (Ed.), *The Alchemy Reader—From Hermes Trimegistus to Isaac Newton*, Cambridge University Press, Cambridge, 2003, pp. 27–28
25. James R. Partington, *A Short History of Chemistry*, Dover Publications, New York, 2015, p. 20
26. Alic, *Op. cit.* (1), p. 41
27. Sue Vilhauer Rosser (ed.), *Women, Science, and Myth: Gender Beliefs from Antiquity to the Present*, ABC—Clio, Cremona, CA, 2008, pp. 11–13
28. António M. Amorim da Costa, *Introdução à História e Filosofia da Ciências*, Europa América, Lisboa, 2002, pp. 78–79
29. A. Khoirul Fata, P. Irpan Fauzan, "Hellenism in Islam: The Influence of Greek in Islamic Scientific Tradition", *Epestimé: Jurnal Pengembangan Ilmu Keislaman*, 13 (2019) 407–432
30. H. J. Blumenthal, "Alexandria as a centre of Greek philosophy in later Classical Antiquity", *Illinois Classical Studies*, 18 (1993) 307–325
31. H. Floris Cohen, "Greek nature-knowledge transplanted: the Islamic world", in *How Modern Science Came into the World: Four Civilizations, One 17th-Century*, Amsterdam University Press, Amsterdam, 2010, pp. 53–76
32. P. G. Maxwell-Stuart, *The Chemical Choir: A History of Alchemy*, Continuum, London, 2008, pp. 1–8
33. T. L. Davis, "Primitive science, the background of early chemistry and alchemy", *J. Chem. Educ.*, 12 (1935) 3–10
34. S. V. Meshel, "Teacher Keng's Heritage—A Survey of Chinese Women Scientists", *J. Chem. Educ.*, 69 (1992) 723–730
35. T. L. Davis, "Ko Hung (Pao P'u Tzu), Chinese Alchemist of the Fourth Century", *J. Chem. Educ.,* 11 (1934) 517–520
36. T. L. Davis, Lu-Ch'Iang Wu, "Ko Hung on the Yellow and the White", *J. Chem. Educ.*, 13 (1936) 215–218
37. P. G. Maxwell-Stuart, *The Chemical Choir: A History of Alchemy*, Continuum, London, 2008, pp. 19–28
38. A. Schwarz, "Indian and Western Alchemy Derivations and Deformations of Patterns", India International Centre Quarterly, 8 (1981) 145–158
39. N. Rout, "Role of Women in Ancient India", *Odisha Review*, (2016) 42–47

2

Nuns and Alchemists

It is quite evident that women, with their intelligence and abilities, have made substantial discoveries in numerous important sciences and techniques, both in the pure sciences—as their writings show—and in the field of techniques, which manual work and crafts prove.

CHRISTINE DE PISAN, *The Book of the City of Ladies* (1405) [1]

Until the twelfth century, several European nuns achieved a high level of erudition, prestige, and authority. Among them, Hildegard of Bingen, a German abbess, stood out as one of the most cultivated figures of her time, displaying extensive knowledge in natural philosophy and medicine. As universities emerged in Europe, gradually supplanting monasteries as centers of culture, they became exclusive domains for men. However, in the early fifteenth century, Christine de Pisan in France initiated a public discourse challenging the prevalent misogyny in society, sparking what would later be known as the "women's quarrel"—a discussion that has extend to our time.

In the sixteenth century, alchemy, introduced to Europe primarily through the influence of Arabs who had settled in the Iberian Peninsula during the eighth century, underwent a division into two main pursuits. One pursued a spiritual and metaphysical path, striving for immortality and the creation of the Philosopher's Stone. The other adopted a more pragmatic approach, focusing on the creation of substances, particularly medicines. Despite being condemned by the Church as a form of witchcraft, especially during the Protestant Reformation, alchemy persisted until the eighteenth century, with significant participation from women. An exemplary figure in this regard

© The Author(s), under exclusive license to Springer Nature Switzerland AG 2024
J. P. André, *Sisters of Prometheus*, https://doi.org/10.1007/978-3-031-57136-7_2

was Dorothea Juliana Wallich, a German pioneer in the study of cobalt chemistry.

2.1   Convents and Universities

The period that ensued after the fall of the Western Roman Empire in 476, marking the start of the Middle Ages according to historians, was defined by a prolonged economic decline in Europe. This decline persisted, particularly until the turn of the millennium, and was further aggravated by invasions, carnage, famine, and disease. Although the term "Dark Ages" had been previously used to describe this period, it is now largely rejected because, while much knowledge from the Greco-Roman civilization and earlier civilizations was indeed lost, significant cultural advancements also occurred. Notable achievements during that time encompassed the development of modern languages, the establishment of the Roman-Germanic civilization, and the proliferation of monasteries.

These monastic institutions played a vital role in promoting literacy and nurturing scholarship by copying and studying surviving fragments of ancient texts [2]. The scientific thought of the Early Middle Ages was predominantly based on these writings. In the meantime, however, the Islamic world had assimilated the teachings of Ancient Greece, adding its own knowledge to them. This fusion led to remarkable advancements, particularly in fields such as philosophy, medicine, mathematics, astronomy, optics, and alchemy. Through the Arab presence in Spain, as well as in southern Italy and Sicily, and facilitated by trade, travel, and the Crusades, this wisdom spread throughout Europe. The Latin translations of works by scholars such as Avicenna (908–1037), a Persian polymath renowned for his contributions to philosophy and medicine, had a significant impact. However, it was the recovery of complete works by influential intellectuals of antiquity, like Aristotle and Galen, that played a crucial role in shaping medieval thought [3].

Monastic institutions indeed had a decisive influence on the rebuilding of Western civilization after the fall of Rome. However, obtaining food and surviving attacks and diseases were far more pressing concerns than receiving an education [4]. If the child happened to be a girl, instruction held even less significance, particularly if she was not of noble birth. Despite the very low literacy rate among women in the Middle Ages, education, especially from the eighth century onward (with the Carolingian Renaissance), was provided in various contexts, among which convents and abbeys stood out. To these ones still added the schools of the cathedrals, as well as those of the castles, courts, and villages, besides private education—in charge of the family, tutors, governesses, and priests.

For women, a life in the service of God was not only an alternative to marriage but also a way of redemption for the Original Sin. Many entered monastic institutions never to leave them; others entered at the age of seven in order to obtain an education before getting married (which was sometimes at the age of 14). In both cases, learning was an important component of convent life. The usual curriculum included reading and writing in Latin, religion, morals, music, painting, etiquette, embroidery, spinning, and weaving. Others included still the Greek language and the seven liberal arts (grammar, dialectics, rhetoric, music, arithmetic, geometry, and astronomy). The frequent invasions and massacres that afflicted those times meant that medicine and surgery were also frequently taught [5].

Although some nuns led isolated and austere lives devoted to religious duties, many of the early medieval monasteries were relatively liberal, offering some comfort and a range of educational and work opportunities. It was common for the more learned ones, normally of noble origin, to become abbesses (those of low social origin were only admitted as servants). The Middle Ages managed to produce a group of women whose erudition surpassed that of the vast majority of their contemporaries, leading them to attain positions of great notoriety and power, including the direction of double monasteries (religious communities with residents of both sexes, in which men and women lived in separate sections but under the same administration). In the seventh century, Hilda of Whitby (c.614–680) stood out, playing a significant role in the Christianization process of Anglo-Saxon England. Considered the first great English female pedagogue, she may have been the most cultured woman in Great Britain before the Norman Conquest [6]. Five centuries later, in Alsace, Herrad of Landsberg (c.1130–1195) and, in Germany, Hildegard of Bingen (c.1098–1179) gained prominence. They would be the last erudite abbesses of the Middle Ages.

Herrad of Landsberg, the author of *Hortus Deliciarum* [*Garden of Delights*], is considered to have created the first female-authored encyclopedia. The work, which she personally illustrated, is a compilation of texts by classical and Arab authors on topics such as religion, history, astronomy, geography, philosophy, natural history, and medicinal botany. Intended for the instruction of novices, it also includes poems and songs.

Hildegard of Bingen, also known as the Sibyl of the Rhine or Prophetess of the Teutons, was a multifaceted figure with various talents. She was a writer, orator, mystic, physician, naturalist, composer, playwright, and illustrator. Born into the aristocracy, she entered the Benedictine Monastery of

Disibodenberg in Germany at the age of eight, where her aunt, Jutta of Sponheim, educated her. At 15, she took religious vows, leading a life of seclusion and asceticism. When her aunt passed away in 1136, Hildegard became the abbess of the congregation. It was through a mystical vision (is it said that she started having these spiritual experiences around the age of five) that she felt she had a divine mission to write and preach. In 1151, with papal permission, she left Disibodenberg and, accompanied by other nuns, went to Rupertsberg near Bingen in the Rhine Valley, where she revitalized the old local monastery.

In her extensive work, Hildegard of Bingen addressed themes of religious and philosophical nature, such as in *Scito Vias Domini* [*Know the Ways of the Lord*], better known as *Scivias*. Additionally, she focused on subjects of natural philosophy and medicine, writing the *Liber Subtilitatum Diversarum Naturarum Creaturarum* [*Book of the Subtleties of the Various Natures of Creation*] between 1151 and 1158. This treatise was divided into two parts after her death: *Physica, Liber Simplicis Medicinae* and *Causae et Curae, Liber Compositae Medicinae*. It presents a holistic view and discusses the causes of diseases and their remedies, which are based on plants, minerals, and animal products. It also delves into the functioning of the human body, considered as the microcosm, and its relationship with the universe, the macrocosm. This work was the first of its kind in the Holy Roman-German Empire [7], remaining a reference until the sixteenth century, particularly in Northern Europe. The content of the work, with its original observations, reveals that Hildegard of Bingen was knowledgeable not only in traditional healing but also in the medicines of Hippocrates and Galen, as well as in the works of the naturalist Pliny the Elder. Many of her remedies share similarities with other herbal medicines, both in Western herbalism and Eastern systems like Ayurveda (see Chap. 1) and traditional Chinese medicine. For example, psyllium is recommended for constipation (as it is still today), aloe for jaundice, and white horehound for coughs, among others [8].

By establishing a close relationship between body and soul, mankind and Nature, and the natural and the supernatural, she advocated that many human diseases were a punishment for the Original Sin. She defended the biological and psychological complementarity of man and woman, attributing the same dignity to both. With regard to the biological conception of the human being, she departed from the Aristotelian explanation (discussed in Chap. 1), attributing to the mother a hitherto unusual importance: it was the heat of her uterus, emanating from the blood, which gave shape to the child. For obvious reasons, Hildegard of Bingen has often been seen as a pioneer of feminism, which, however, can be contested with the argument that she never aimed at female emancipation or questioned the role attributed to women in society [9–11].

While Hildegard of Bingen's work represents the pinnacle of female scientific writing in the medieval period, it is perplexing how she managed to maintain her independence from the Church and acquire such authority, particularly considering that her active period coincided with the Papal Reformation, which began in 1050 and lasted until the thirteenth century. In fact, she did not only maintain an intense correspondence with several popes and sovereigns, who asked her for advice on spiritual and secular topics, as she also preached Christian doctrine in public throughout several regions of Germany. To this contributed her immense wisdom and, above all, the fact that she never questioned the authority of the Church, always respecting

Fig. 2.1 Hildegarde of Bingen, bestowed with divine inspiration, imparts it to a monk. Illustration *from Scivias* (1141–1152)

orthodoxy. Moreover, she attributed all her works to God and not to herself, claiming to be nothing more than a mere messenger or passive recipient of knowledge from above, which is represented in illuminations such as the one in Fig. 2.1. She herself nurtured the idea of being an *indocta* (ignorant) and *paupercula* (poor) woman [12].

The Papal Reformation, with Gregory VII playing a particularly prominent role during his pontificate from 1073 to 1085, encompassed a series of measures. These measures had two primary objectives: a return to the roots of Christianity and, equally importantly, the affirmation of the Pope's authority over feudal powers. To combat the prevailing clerical degeneration, the obligation of celibacy was strengthened, eventually leading to an increase in misogyny and further discrediting women. Double monasteries were abolished, the autonomy of women's religious communities was restricted, and nuns were subjected to greater seclusion. On the other hand, the growing importance of the cathedral schools, which were almost exclusively the primary institutions for teaching Latin and denied women access, played a crucial role in accentuating the disparity in educational opportunities between men and women. Consequently, this period marked the decline of an era when several abbesses were renowned for their erudition and social status [13].

The emergence of universities, which replaced the monasteries as the main intellectual centers, brought with them a diverse range of studies in theology, philosophy, logic, mathematics, astronomy, law, and medicine. These universities constituted a powerful and intellectually revitalizing force of enormous importance for Europe. However, they were limited to a male elite and heavily influenced by a strong clerical character. Their androcentrism was legitimized by a discourse based on patristic literature and the philosophical and medicinal texts of the Aristotelian-Galenic tradition, which portrayed women as "imperfect males." With the exception of the universities in the Italian Peninsula, particularly the University of Bologna (1088)—the first in Europe—these institutions were directly associated with the Church. The University of Paris (1170), founded from the school of Notre Dame Cathedral, serves as a prime example of this connection. However, apart from a few specific cases, only Italian universities have not completely banned access to women. For instance, Bettisia Gozzadini, a doctor in law, benefited from this opportunity and taught in Bologna during the thirteenth century, despite having to wear male clothing for that purpose. Interestingly, Italian universities were also less strict in terms of mandatory celibacy for their instructors [14]. Before that time, some women had already been admitted as masters and disciples to the Medical School of Salerno, situated in the south of Naples. This institution, with roots in the Greek and Arab legacy, stood as the most important center of medical knowledge in Europe between the tenth and thirteenth centuries. Among these women, Trotula of Salerno stood out in the eleventh century, although little is known about her [15]. Several treatises on medicine are attributed to her, the most important being *De Passionibus Mulierum Curan Dorum* [*On the Healing of Women's Sufferings*], known as *Trotula Maior*, which mainly focuses on gynecology and obstetrics. Additionally, *Ornatu Mulierum* [*Women's Ornamentation*], or *Trotula Minor*, became the first book on female body esthetics written by a woman practicing medicine. This work explores cosmetics and skin diseases, offering recipes for preparations based on plant, animal, and mineral products serving both esthetic and medicinal purposes. (Many of the plant-based ingredients mentioned in her work are still used in various cosmetics today [16].) In the subsequent centuries, women from different European countries, with varying degrees of affiliation with educational institutions and professional corporations, occasionally engaged in surgical and medical practices, especially in the field of obstetrics [17].

2.2   The Value of Women

At the onset of the fifteenth century in France, a literary and philosophical debate concerning the capabilities of women and their societal status unfolded. This discussion, famously known as *querelle des femmes* ("women's quarrel"), persisted until the twentieth century [18], and it originated from the latent conflict between the enthralled depiction of women sung by the troubadours and the misogynistic conception presented by both the Church and the short narratives in verse known as *fabliaux*. These popular tales, characterized by humorous and sarcastic content, ruthlessly mocked and demeaned women. On one hand, the Church perceived women as the incarnation of Eve and, consequently, the source of all evil—a notion reinforced by the *fabliaux*, which portrayed them as libertine and dangerous beings, ready to betray their husbands if given the chance. On the other hand, the troubadours held a contrasting view, celebrating the nobility of women and arguing that true love could only be found outside of marriage, especially in an era when many marriages were merely political and economic alliances between families or states. The *Roman de la Rose* [*The Romance of the Rose*], one of the most famous literary works of medieval France, exemplarily illustrates these two contrasting conceptions of women—the depraved and the passionate ones. This work, an allegorical poem in two parts about a love dream, incidentally, is also brimming with references to alchemy. Its first part was written in the 1230s by

Guillaume de Lorris, and the second part was authored between 1275 and 1280 by Jean de Meung. Little is known of the first author, an aristocrat who celebrated courtly love, while Meung, a bourgeois who fiercely attacked the female sex in the manner of the *fabliaux*, is known to have graduated from the University of Paris. The *Roman de la Rose* precisely incited the indignation of Christine de Pisan (1363–*c*.1430). By 1399, she had already written *Épistre au Dieu d'Amours* [*Epistle to the God of Loves*], an allegorical poem defending the dignity of women, arguing that the lack of virtue in some did not make all women the malevolent beings portrayed by Meung. It was within this work that Christine de Pisan, for the first time, extolled women's intellectual capacities, a theme she systematically developed thereafter. In the literary circles of Paris, the statesman, humanist, and writer Jean de Montreuil, in turn, circulated a now-lost treatise championing the *Roman de la Rose*, to which Christine de Pisan replied, in 1401, with an open letter denouncing the immorality, misogyny, and obscenity of the work. This marked the establishment of the debate that would evolve into the embryo of feminism [19]. With the appearance of Jean de Gerson, chancellor theologian at the University of Paris and author of *Tractatus Contra Romantium de Rosa* [*Treatise Against the Romance of the Rose*, 1402], Christine de Pisan found an ally. Gerson referred to her as a "remarkable and virile woman" [20]. On the opposing side, along with Jean de Montreuil, were the brothers Gontier and Pierre Col. The first, the king's secretary, demanded she retract her words, while the second, a canon, accused her of lacking the modesty expected of a woman.

Born in Venice in 1364, Christine de Pisan (Fig. 2.2) was the daughter of Tommaso da Pizzano. Due to her father's appointment as Charles V's physician, privy adviser, astrologer, and alchemist, the family moved to Paris while she was still a child. From her father, she inherited a thirst for knowledge and a taste for studies. Tommaso da Pizzano, an open-minded man, wished to provide his daughter with a more extensive education, recognizing her refined intelligence. However, the customs of the time did not permit such opportunities for women. Nevertheless, he taught her Latin and philosophy, subjects that were rarely part of female education during that era.

At the age of 15, Christine married a notary and secretary to the king, who succumbed to the plague in 1389. As her father had already passed away, she saw herself, at the age of 25, with no means of subsistence and the responsibility of supporting three children, two brothers, and her mother. In this challenging situation, she made a life-altering decision to pursue writing as a means of livelihood. Throughout her literary career, she wrote on various subjects and in different genres, including epistles, ballads, poems, moral treatises, and educational works. She also served as the official biographer of

Fig. 2.2 *Christine de Pisan* teaching four men. Miniature from *The Queen's Manuscript,* *c.*1410–1414. British Library

Charles V. Later in life, at the age of 54, she entered a convent where she resided until her passing.

It was in works such as the *Livre de la Cité des Dames* [*The Book of the City of Ladies*, 1405] and the *Livre des Trois Vertus* [*The Book of Three Virtues*, 1405] that Christine de Pisan, the first female author in Europe to make a living from writing, widely exposed her ideas about the value and role of women in society [21]. These two books, about an imaginary city inhabited only by women, where everyone had the right to education and the nobility of the spirit counted more than that of the cradle, contain dialogues with three allegorical figures—the feminine virtues—who are ready to help in the construction and population of the city: Reason, Rectitude, and Justice. The use of characters of this type was common among medieval (male) authors, especially in instructional books considered true mirrors for behavior. In *The Book of the City of Ladies*, the narrator asks Reason if God ever wanted to ennoble women's minds with something as elevated as science (knowledge). "Has He given them enough intelligence to do so?" And she goes on, recalling the constant male claim that women learn little. To this, Reason responds: "The truth is the opposite of what they say," adding that "if it were customary to send girls to school, as it is with boys, they could learn and understand the subtleties of all the arts and sciences as well as them"; women had actually "freer and more lively minds" than men [22]. In her book, Christine de Pisan evokes over one hundred and fifty female figures from history, religion, legend, and mythology. She drew inspiration from Boccaccio's *De Mulieribus Claris* [*Famous Women*, 1374], in which the Italian writer compiled biographies of renowned women (not all of them for the best reasons).

The contribution of Jean de Gerson, who temporarily calmed the "women's quarrel," was later reinforced after Christine de Pisan had already passed. This reinforcement came in the form of *Le Champion des Dames* [*The Champion of Women*], a poem written in 1442 by Martin Le Franc, who served as the provost of Lausanne. This work vigorously defended women, emphasizing female rationality and challenging the prevailing orthodoxy. In *Le Champion des Dames*, Le Franc expressed his opposition both to the views presented by Meung in *The Romance of the Rose* and those conveyed by the Church Fathers in the past. However, almost a century later, already in the midst of the Renaissance, the publication of *Controverses des Sexes Masculine et Feminine* [*Controversies of the Male and Female Sexes*, 1534], a long poem by Gratien du Pont, was another virulent diatribe against women, reactivating the old feud. Nevertheless, the sixteenth century was prodigious in works praising women and advocating their access to education, of which the celebrated treatise *De Nobilitate et Praecellentia Foeminei Sexus* [*Declamation on the Nobility and Preeminence of the Female Sex*, 1509], written by Cornelius Agrippa, from Germany, and later translated into French in 1530, is a notable example.

An important female person in the culture of the French Renaissance was Marguerite d'Angoulême (1492–1549), sister of Francis I and herself Queen Consort of Navarre by marriage to Henry II. At her court in Nérac, visited by intellectuals, literati, and poets, figures such as the priest and reformer John Calvin and the physician, writer, and priest François Rabelais stood out. In addition to speaking several European languages, Marguerite d'Angoulême studied Latin, Greek, philosophy and theology, as well as mathematics, medicine, geography, and cosmography, having also devoted herself to writing. Similar to Christine de Pisan, she used her literary voice to denounce the difficulties that women faced in a world dominated by men. Her best-known work is *L'Heptameron* (published posthumously in 1558), which was modeled on Boccaccio's *Decameron*, written about a century earlier [23, 24].

Given the growing number of works by women in aristocratic and bourgeois circles, the character of the *femme docte* (learned woman) was praised, and satirized, by various authors. This new reality is evident in Rabelais's comic novel *Gargantua* (1534), in which the protagonist, the giant Gargantua, enthusiastically describes to his son, Pantagruel, the good times they were going through, in which "women and girls longed for the heavenly praise and manna of good doctrine [erudition]" [25, 26].

2.3   *De Transmutationibus Metallorum*

The secrets of Alexandrian alchemy as the art of transforming base metals into gold and silver (see Chap. 1), along with artisan techniques for imitating these precious metals and stones, were transmitted to the Byzantine Empire [27]. One of its empresses, Zoë Porphyrogenita (*c.*978–1050), became famous both for her active political and love life and for having a laboratory where she created perfumes and ointments [28].

Constantinople, which served as the capital of the Eastern Roman Empire from 330 onward, played a crucial role in the dissemination of alchemical knowledge to the European continent, a process further facilitated by the Crusades. Additionally, the Arab presence in the Iberian Peninsula, starting in 711 and two centuries later in Sicily, contributed significantly to the spread of alchemy and the preservation of the culture of Classical Antiquity. Many Islamic works were translated into Latin by members of religious orders. In the realm of alchemy, a pivotal moment occurred in 1144 when the English monk and Arabist Robert of Chester completed the translation of the *Liber de Compositione Alchemiae* [*Book of the Composition of Alchemy*]. However, the practice of this discipline faced a setback in 1317 when Pope John XXII issued the papal bull *Spondent Pariter*, which prohibited it. This decree, while rooted more in moral concerns than religious ones, aimed to combat the fraudulent activities of numerous self-proclaimed alchemists who exploited the vulnerable and impoverished. Interestingly, Pope John XXII himself authored *Ars Transmutatoria Metallorum* [*The Art of Transmuting Metals*] [29].

Between the late Middle Ages and the end of the seventeenth century, European alchemy underwent an extensive process of redefinition, both in its internal structure and in its interaction with other areas of knowledge. It transformed from a marginal domain with a strong mystical dimension, centered on metallurgy, mineralogy, and certain chemical processes (particularly those associated with the production of products such as pigments, dyes, ammunition, ceramics, and glass), into a branch of knowledge that laid the foundations for a new therapeutic practice and a theory of matter. Gradually, it sought to become integrated into natural philosophy, as will be discussed in Chap. 4 [30, 31].

It was precisely in the context of preparing chemical remedies, following the path of Paracelsian iatrochemistry, that alchemy reached its zenith during the sixteenth and seventeenth centuries. Iatrochemistry, which was alchemy in service of medicine, had been championed by Philippus Aureolus Theophrastus Bombastus von Hohenheim (1493–1541), a Swiss better known as Paracelsus. In his early years, he worked as an analyst in mines and pursued medical studies in Basel, Vienna, and Ferrara. However, he eventually rejected the teachings of Galen and Avicenna. Defender of the idea that the four classical elements were presented in a triad of fundamental principles—sulfur, mercury, and salt, known as the *tria prima*—and that all existing substances were formed from their combination in different proportions, Paracelsus not only reconfigured alchemy in its philosophical-spiritual essence but also applied it to medicine [32]. Regarding the former, he believed that alchemy was the key to a truly Christian interpretation of Nature. As a doctor, he utilized metallic salts, particularly mercury, to treat syphilis. However, he also believed in the active principles of plants and therefore employed extracts, tinctures, and essences of vegetable origin in his medical practice [33].

The line of separation between alchemy and chemistry remained blurred until late, with the terms "alchemy" and "chemistry" being used interchangeably until the end of the seventeenth century. It was not before the third decade of the eighteenth century that "alchemy" began to exclusively refer to the discipline aimed at the transmutation of metals. The English language has the advantage of having the term "chymistry," which encompasses both alchemy and early chemistry in an undifferentiated way. This term is particularly suitable for the period between the first quarter of the sixteenth century and the end of the seventeenth century [34].

Early on, European women were drawn to the alluring world of alchemy, and this fascination would endure for centuries. The association of the female figure with alchemy, along with the portrayal of women within the alchemical universe, can be found in both manuscripts and printed works, as well as in paintings from various periods. Women are frequently depicted representing concepts traditionally associated with feminine traits, such as stars (the moon and Venus), and actively participating in laboratory processes, particularly those involving operations without heating (Aristotle believed women were cold) or those conducted in an aqueous medium or requiring volatilization.

Quelle: Deutsche Fotothek

Fig. 2.3 Representation of alchemy as a woman. Woodcut from 1574

An example of such representation can be found in *Quinta Essentia*, a book by the Swiss alchemist Leonhard Thurneysser, where alchemy is allegorically depicted by an image of a woman (Fig. 2.3). Positioned between two furnaces and flanked by the sun and the moon, she holds an alchemical flask and tweezers.

In the famous treatise *Atalanta Fugiens* [*Atalanta Fleeing*, 1617], written by the German alchemist Michael Maier, laboratory alchemical practice is compared to the work of a woman in the kitchen. One of its illustrations (Fig. 2.4)

Fig. 2.4 Emblema XXII from Michael Maier's *Atalanta Fugiens* (1618): *Plumbo habito candido fac opus mulierum, hoc est, COQUE*. Courtesy of Science History Institute

portrays a pregnant woman standing next to pots on a fire. The corresponding caption reads "Having acquired white lead [basic lead carbonate], do the work of women, that is, cook." Pregnancy in this image symbolizes the incubation process in the alchemist's vessel, which was believed to be necessary to achieve the Philosopher's Stone.

Women are also often represented alongside their alchemist husbands. In one of the best-known prints, *The Alchemist* (Fig. 2.5), based on a drawing by Pieter Bruegel the Elder and dated from *c*.1558, a male figure (on the left) is seated by a furnace, attempting in vain to transmute lead into gold. His failure drags the entire family into misery, as depicted in the scene visible through the window. In the center, his wife searches her pocket for the last gold coin, while the neglected children rummage through a cupboard in search of food. On the right side, a nun examines several opened books.

In turn, in *The Alchemist and His Wife in the Laboratory* (Fig. 2.6), a painting from the Flemish school of the seventeenth century by David Ryckaert

Fig. 2.5 The Alchemist (*c.*1558). Engraving by Philips Galle after Pieter Bruegel the Elder. Metropolitan Museum of Art, New York

III, an artist who, in several works, expressed his fascination with this theme, portrays the alchemist's wife in a manner distinct from the conventional roles typically depicted (such as the wife responsible for the household chores, tending to the children, or lamenting her husband's unsuccessful efforts). In this canvas, the woman at the center holds a thick book and a pair of glasses. The nature of the book raises intriguing questions: is it the Bible, suggesting a moralizing attitude toward her husband, or is it simply a compendium of alchemical knowledge, implying her efforts to assist him [35]?

It was indeed through association with her husband that the name of a French medieval alchemist came to us: Pernelle Lethas (*c.*1330–1397). She entered into her third marriage in 1368, this time to Nicolas Flamel, who was a public clerk, copyist, and bookseller in Paris. Flamel (*c.*1330–1418) had acquired in 1357 an ancient book that was allegedly revealed to him by an angel during a dream. The book, bound in copper and with 21 leaves made of tree bark, contained a reference to Abraham and was mostly composed of symbols and figures whose meanings eluded the clerk. Despite his futile attempts to decipher them, in 1378, he embarked on a journey to Spain in the hope of finding a knowledgeable Jew who could assist him. (In France, Philip

Fig. 2.6 *The Alchemist and His Wife in the Laboratory.* Copy from circa 1928 of the original by David Ryckaert III (1648). Courtesy of the Science History Institute

IV had ordered the expulsion of the Jews in 1306, and subsequent decades witnessed ongoing persecution of Torah followers.) Two years later, in Leon, he encountered Master Canchez, an elderly merchant from Bologna and a Kabbalah specialist. Canchez, actually a Jewish physician who had converted to Christianity, immediately recognized the drawings reproduced by Flamel, asserting that they originated from the *Asch Mezareph*, a long-lost book. He volunteered to accompany the clerk back to Paris, where they could analyze the work in question. However, due to his advanced age, he never reached their intended destination, passing away in Orléans after sailing up the Loire.

Despite this setback, Canchez's preliminary decipherments proved decisive for Flamel. After a few more years of efforts, prayers, and reflections, he finally managed to understand the alchemical message of the book. In January 1382, he reported successfully using the Philosopher's Stone to transmute mercury into silver (*l'ouevre au blanc*). In his account, he gave great credit to his wife's participation in the laboratory work, always referring to her affectionately and stating that she knew as much about alchemy as he did. Continuing their pursuit, the couple claimed in April to have transmuted the same metal into gold (*l'oeuvre au rouge*), which they alleged was more malleable than common gold. In the first case, it may have involved the reduction of silver ores followed by the distillation of mercury; in the second, it may have been an identical process starting from gold ores.

Anyway, the story goes that it was with the wealth thus obtained that they made generous donations for religious and charitable works. Over time, there have been many legends surrounding the couple, and some have believed that they were both still alive in the seventeenth century. There were also several alchemical treatises attributed to Flamel, the most famous of which was *Le Livre des Figures Hiéroglyphiques de Nicolas Flamel* [*The Book of Hieroglyphic Figures by Nicolas Flamel*], published in Paris in 1612 by Arnauld de la Chevallerie (and translated into English and published in London in 1624). This is where the story of the mysterious book revealed to Flamel by an angel is told. Today, the text is believed to have been written between the late sixteenth- and early seventeenth centuries. However, in favor of Flamel's real existence, there is historical evidence such as his tombstone, with the inscription of his name. Currently on display at the Musée de Cluny in Paris, it proceeded from the ancient church of Saint-Jacques-de-la-Boucherie, which the clerk is said to have financed in 1389. All indicates that he lived in 51 Rue de Montmorency, one of the oldest Paris buildings. Pernelle, in turn, is represented in a figure of *Le Livre des Figures Hiéroglyphiques* that reproduces the images carved in the arch of the cemetery of the Church of the Innocents (one of several that the couple had commissioned, but which has not survived).

The names of Flamel and Pernelle have been preserved in two streets of the French capital, and the clerk alchemist appears in recent works of fiction such as *The Da Vinci Code*, by Dan Brown, and *Harry Potter and the Philosopher's Stone*, by J. K. Rowling. According to the former, Flamel was one of the Grand Masters of the Priory of Sion, and the latter, besides attributing the creation of the Philosopher's Stone to him, tells that he lived with his wife for over 600 years.

Anne, Princess of Denmark and Norway (1532–1585), and her husband, Augustus I, Elector of Saxony, also devoted themselves to the practice of alchemy. In 1578, they claimed to have produced "three ounces of gold from six ounces of silver" with the help of a Swiss alchemist named Sebald. The Electress of Saxony was also dedicated to the preparation of pharmaceutical products, reflecting the interest that Paracelsian iatrochemistry was beginning to arouse. For that purpose, in Annaburg (a city named in her honor), she had at her disposal the best laboratories in sixteenth-century Germany. Several women working for the Electress gathered all kinds of herbs, berries, leaves, roots, and flowers in her gardens and in the neighboring fields and forests.

These botanical resources were then dried and stored. To these ingredients were added, among others, pulverized human bones, moss grown on skulls, human and dog fat, ox bile, donkey milk, deer and goat blood, as well as the much-desired unicorn horn powder. Anne of Denmark, who collected recipes from all kinds of healers of the time, using these and other ingredients (which, however bizarre, could be found in many dispensaries of the sixteenth century), composed ointments, syrups, electuaries, and distilled spirits. Her correspondence shows that she was in contact with important physicians and alchemists and that it was due to her mother's influence that, from an early age, she became interested in the production of pharmaceutical preparations. She acquired knowledge of distillation and alchemical practices from the court physician Paul Luther (1533–1593), the son of the renowned German reformer [36].

Bianca Cappello (1548–1587), from Venice, the second wife of Francesco I de' Medici (great-grandson of Caterina Sforza—see Chap. 3), is also believed to have practiced alchemy with her husband. Strangely, both died on the same day, after a long agony, according to a clinical condition compatible with arsenic trioxide intoxication [37].

In seventeenth-century France, there has been another couple deeply associated with alchemy: Martine de Bertereau (c.1578–c.1642), the Baroness de Beausoleil, and her husband, Jean du Châtelet, the Baron de Beausoleil and d'Auffenbach. Together, they dedicated themselves to exploring mineral deposits, not only in France (at the service of Henry IV, who wanted to relaunch the economy, devastated by the religious wars) but also in various European countries and even in Bolivia, in the famous silver mines of Potosí.

A cultured and intelligent woman who possessed a wide range of knowledge, Martine de Bertereau was fluent in Latin and, in addition to her native tongue, could converse in German, Italian, and Spanish. The Baroness de Beausoleil acquired from her husband the essential teachings for mining, as he believed that this pursuit required a comprehensive set of skills, including astrology, architecture, geometry, arithmetic, hydraulics, mechanics, mineralogy, alchemy, pyrotechnics, botany, as well as expertise in legislation and theology [38]. As a woman, one can consider her a pioneer in fields that today correspond to geology and mining engineering.

Martine de Bertereau hailed from a noble family in the region of Touraine or Berry in central France. In 1601, she married Jean de Châtelet, who originated from Brabant in the Spanish Netherlands. It was through their union that they embarked on their collaborative endeavors. However, their unconventional activities and methods raised concerns among the provincial clergy, who feared that they might be involved in magical practices. Consequently, an investigation was launched to find evidence against them. Despite the search yielding no incriminating evidence, the couple was compelled to leave France. However, with the ascent of Louis XIII to the throne in 1610, they were able to return and resume their work.

Both left written work with a strong alchemical character. *Diorismus Verae Philosophiae: De Materia Prima Lapidis* [*Definition of the True Philosophy: On the First Matter of the Stone*, 1627], a small treatise on alchemy, was authored by the baron. In turn, the written legacy of the baroness includes the book *Véritable Déclaration de la Découverte des Mines et Minières de France* [*True Declaration of the Discovery of Mines and Mining of France*, 1632] and *La Restitution de Pluton* [*The Restitution of Pluto*, 1640]. The latter is a poem dedicated to Cardinal de Richelieu which was nothing more than an appeal to be paid for the work of national interest that they had carried out at their own expense. In it, the author recounts her life and that of her husband, saying that they were both devoted to mining and mineralogy. She also mentions the methods they used—several of them associated with astrology and alchemy—as well as the equipment, from compasses to astrolabes. She states that her wish was to return to the king and his people all the riches buried in the depths of the Earth, and she alludes to the little elves that could be found in such concealed places [39]. Although it is not known what happened, Jean de Châtelet was arrested and imprisoned in the Bastille shortly after the publication of this work. Martine, accompanied by her eldest daughter, had the same fate at the Château de Vincennes. Both died in prison, she in 1642 and he 3 years later [40].

Similarly, in *Vitulus Aureus* [*The Golden Calf*, 1667], an alchemical work roughly contemporary with the Barons of Beausoleil, a husband (the author) expresses his pride in his wife's achievements. He is the German-born Dutch physician and alchemist Johann Friedrich Helvetius (1630–1709), who reports that his wife had prepared gold from lead, for which she had a small stone. Allegedly, the transmutation, "which yielded the finest and best gold," took no more than a quarter of an hour [41].

At the end of the eighteenth century, another couple of alchemists stood out in Paris: Sabine Stuart de Chevalier and Claude Chevalier. A military doctor and author of several medical works, he was considered to have extraordinary therapeutic abilities, attributed to the medicines prepared by himself from ingredients of botanical origin. Claude Chevalier sought an elixir that would cure all ills, attainable with the Philosopher's Stone, a subject that he revealed in *L'Existence de la Pierre Merveilleuse des Philosophes, Prouvée par des Faits Incontestables* [*The Existence of the Wonderful Philosopher's Stone, Proved by Indisputable Facts*, 1765]. His wife, who was initiated by him in the discipline, was responsible for the two-volumes of *Discours Philosophique sur les Trois Principes, Animal, Végétal et Minéral, ou la Clef du Sanctuaire Philosophique* [*Philosophical Discourse on the Three Principles, Animal, Vegetable and Mineral, or the Key to the Philosophical Sanctuary*, 1781]. The manuscript was submitted for approval to Pierre-Joseph Macquer (1718–1784), professor of chemistry and member of the Academy of Sciences (opposite of Antoine Lavoisier). Despite not having great admiration for alchemists, Macquer, in his famous *Dictionnaire de Chimie* [*Dictionary of Chemistry*, 1766], still regarded fire, air, water, and earth as the "simple bodies" of which all "compound bodies" were made [42]. It is thus not surprising that he approved the publication, noting that, despite being "of pure alchemy," he had found nothing in it that "should prevent its printing" [43].

The alchemy of Sabine Stuart was not, however, that of transmuting metals into gold. In the preface of her work, she mentions that her mission was to lead readers "to the Garden of the Hesperides, to harvest there the golden apple and distribute it to the unfortunate who need help," specifying that the "the much-desired golden apple is the tree of life, the universal medicine or the drinkable gold that so quickly cures the most desperate diseases and prolongs life." In one of the two illustrations of the work, Sabine saw herself in the female character who, with the Garden of the Hesperides in the background, crowns the monk and medieval alchemist Basilius Valentinus, famous for his treatise *Currus Triumphalis Antimonii* [*The Triumphal Chariot of Antimony*, 1646]. Next to him, over a furnace, there is a glass balloon containing two homunculi [44], one male and one female (Fig. 2.7). In the opinion of historian Jean-Pierre Poirier, the treatise of Sabine de Chevalier—a set of alchemical recipes interspersed with not very brilliant philosophical reflections—does not quite measure up to what the author had set out to do [45]. And to top it off, the great chemical revolution was about to break out at the time of its publication (see Chap. 5).

There are also records of several European women who, between the sixteenth and eighteenth centuries, dedicated themselves to alchemical practices without the help or company of a husband. In Paris, Marie Le Jars de Gournay (1565–1645) stood out as a figure of public recognition for her literary work, especially for editing the famous *Essays* by Michel de Montaigne (1533–1592). Coming from the French gentry, Marie de Gournay (Fig. 2.8) was a child eager for knowledge. She already mastered Latin in her adolescence, which she learned on her own. At 18, she had access to one of the first editions of Montaigne's essays, which immediately seduced her. Five years later, in 1588, during a trip to Paris with her mother, she managed to send a note to the renowned philosopher and essayist. The message produced results without

Fig. 2.7 Engraving from Sabine Stuart de Chevalier's *Discours Philosophique*

delay: they met the very next day. From that sudden acquaintance, a great friendship was born, or even something more. The young woman became his secretary, assistant, translator of quotations from classical authors, inspirational muse, and, posthumously, his editor and commentator.

Montaigne referred to her in his work as his "adoptive daughter" (*fille d'alliance*), admitting "having no eyes for anyone else in the world." The great essayist, who had once considered women superficial and futile, unexpectedly, in the course of his acquaintance with Marie, would end up observing that "males and females" did not have great differences between them, even

Fig. 2.8 Marie de Gournay. Lithograph from nineteenth century

admitting that women "were suitable for all studies, exercises, and jobs." Marie, in turn, in works such as *Egalité des Hommes et des Femmes* [*Equality of Men and Women*, 1622] and *Grief des Dames* [*Complaint of the Ladies*, 1626], defended the equality of the sexes, advocating for women's access to education and to public offices.

   In 1591, having settled in Paris and determined to make a living from her writing, Marie de Gournay undoubtedly drew inspiration from the example of Christine de Pisan. However, she soon realized that achieving financial stability through her literary works was not an easy task. The monetary constraints she faced may have played a role in her decision to dedicate herself to alchemy, a pursuit that she embraced from 1597 onward, despite facing a barrage of criticism. She was well aware that her engagement in alchemy was perceived by many as "perfect madness," as she herself attested:

I really do not know whether this science is truly madness, as some say. However, I am well aware that our most illustrious and recent kings have embraced it, along with the most skilled and qualified people in France. Furthermore, it would be reckless to firmly assert that alchemy is insane, given that the essence of its secrets and faculties remains unknown to us [46]

The person who instilled in Marie de Gournay the foundations of alchemical thinking and provided her with access to a furnace in a glass (or pottery) factory, not far from her home, was Jean d'Espagnet, a friend of the Montaigne family. He became president of the Bordeaux Parliament in 1602 and was also an alchemist with published work. In 1609, under the pseudonym of Chevalier Impérial, he published *Le Miroir des Alchimistes* [*The Mirror of the Alchemists*], a small compendium that included the section "Instructions to Ladies to Be Beautiful and Improved Henceforth Without Using More Poisonous Make-up." In this section, d'Espagnet advised his female readers, among other things, to replace toxic face paints, particularly *blanc d'Espagne* (basic lead carbonate), with products such as *huile d'argent*, an oil that allegedly could also stop certain diseases, or talcum powder, which he taught how to prepare [31].

As was the case with any alchemist of that time, Marie de Gournay's main aim was the quest for the Philosopher's Stone. For this purpose, she utilized various metals (mercury, lead, antimony), which were introduced into the "philosophical egg," a small long-neck glass flask sealed with the "seal of Hermes" (see Chap. 1). The "philosophical egg" was then placed in the athanor, a furnace with three levels also known as the "cosmic oven." The lower level contained the fire, the intermediate level housed the "egg" on a bed of sand, and the upper one, shaped like a dome, reflected the heat. It was believed that after a long period of heating, this operation would result in the Philosopher's Stone, appearing in the form of a red powder. Despite never having achieved her goal, Marie did gain a good command of the laboratory techniques of alchemical operations (calcination, fixation, dissolution, digestion, distillation, sublimation, fermentation, etc.). From then on, she found comfort in the idea that the most important thing was not to find the solution to the riddles that alchemists used "to hide their secrets through nebulous language." What interested her above all was understanding the meaning of the phenomena she observed: arsenic vapors whitened copper; mercury sprinkled on molten sulfur produced a black product (metacinnabar), which, when heated, volatilized without altering its composition, acquiring, however, a beautiful red color (cinnabar); copper disappeared when a strong acid was added to it, giving rise to a transparent green liquor; and from this liquor, when a piece of iron was dipped into it, it was possible to recover the original copper, while the iron was no longer visible. In *Peinture de Mœurs* [*Painting of Costumes*], a self-portrait in verse dedicated to d'Espagnet, published in 1634, Marie alludes to the "mad ramblings" of alchemy, which he had already denounced in his *The Mirror of the Alchemists*:

> Alchemy is my home, but not its crazy ramblings:
> To deceive, to spend a lot, to believe in the art without doubting,
> Waiting for a sea of gold, announcing it without end.
> I didn't deceive anyone, I didn't spend much,
> I crave little, I say less, I hope without belief.
> I neither deceive nor cheat. I don't believe the fake.
> I am all true, and in full faith [47]

The self-defense attitude toward the practice of alchemy is a constant throughout her work, which should not be surprising, considering that the practice of this discipline was condemned by the Parisian authorities in 1624. Additionally, 46 years earlier, the Faculty of Theology of Paris had censored the theses presented by Paracelsus.

Marie de Gournay's work as editor and annotator of the 1595 edition of Montaigne's *Essays*, and, not least, her own writing, won her respect and admiration both at home and abroad. She corresponded with personalities of letters and politics across Europe. She was received at the court of Henry IV and obtained the protection of Cardinal de Richelieu. At her house on Rue de l'Arbre, in the company of her three female cats—*Donzelle*, *Minette*, and *Piaillon*—she was visited by members of the Academy and aristocrats. With time, and with her position of literary authority well established, she became embroiled in the intricacies of politics, taking up the idea of equality between men and women, originally advocated by Christine de Pisan. Arguing that the subordination of women was not willed by God or Nature but simply imposed by the crude arrogance of men, she saw in the pursuit of knowledge an opportunity for women to advance their cause [31, 48].

There are alchemical manuscripts written by women during the first half of the seventeenth century which originated in France or Switzerland and followed the work of Joseph du Chesne (c. 1544–1609). Also known as Quercitan, the latter was the most important figure of paracelsianism in France. The primary focus of these women's writings, including those of Madame de la Martinville, was on the processes of producing gold and elixirs, as well as on spiritual and philosophical questions. This distinguishes them from other texts written by women that are associated with daily and domestic tasks, such as the preparation of perfumes or pigments, as explored in Chap. 3 [49, 50].

In the beginning of the eighteenth century, the important work of the German alchemist Dorothea Juliana Wallich (1657–1725) stands out. Under the pseudonym of D. I. W., and driven by the search for the Philosopher's Stone, she published three books in 1705 and 1706, in which she brought together the results of her investigation of a cobalt ore, which she refers to as "minera." Although the discovery of cobalt is attributed to the Swedish chemist Georg Brandt (1694–1768), Dorothea Wallich seems to have been a pioneer in the observation of some chemical reactions of this element and the thermochromism of some of its compounds.

Born in 1657 in Weimar, she was the daughter of a tax collector for the Duke of Saxe-Weimar, and she married Johann Wallich, secretary to the ducal court, at the age of 16. Dorothea's interest in alchemy began around 1685. The little that is known of her life is thanks to the physician and alchemist Georg Ernst Stahl (1659–1734), the author of the phlogiston theory (see Chap. 5). Stahl, who lived in Weimar between 1687 and 1694, where he served as the physician of the Duke, knew Dorothea Wallich closely. She even was the godmother to one of his daughters. According to him, Wallich "no doubt had more experience in [al]chemical things than a large number of academics and intellectuals" [51]. Her three books on the "minera," which she also referred to as "Electrum minerale immaturum," "Magnesia," "Markasita plumbea," or "Wismuth," as she believed it was necessary to obtain the Philosopher's Stone, were the result of several laboratory investigations and a careful study of the existing literature. The ore in question was probably native bismuth accompanied by safflorite ((Co,Fe)As$_2$) and/or skutterudite (CoAs$_3$), which were abundant in the Schneeberg mines in Saxony, where, according to Stahl, she had conducted prospecting. Once the bismuth component of that ore was removed (by smelting), the remaining part was commonly used in the production of cobalt blue glass.

The historian Alexander Kraft analyzed Wallich's books and, using current chemical nomenclature, provided an interpretation of the experiments she conducted with the "minera" [52]. It is in her third book, *Key to the Cabinet of the Secret Treasure of Nature, for the Search and Discovery of the Philosopher's Stone* [53], published in Leipzig in 1706 that she describes in detail the various chemical tests carried out. By adding sodium chloride to the pink solution produced by reacting the ore with nitric acid, Wallich likely obtained a solution of cobalt(II) chloride, from which, by evaporation, she must have obtained the corresponding salt, CoCl$_2$. She also observed the thermochromism of this compound and its aqueous solutions. When heating the solid cobalt(II) chloride, she found that its color changed from pink to green, passing through blue-purple and sky blue. Upon cooling, the reverse process occurred, returning to pink [54].

Wallich also made the "minera" to react with other substances, namely, with citric acid, ammonium chloride, potassium nitrate, sodium sulfate, antimony sulfide, mercury(II) chloride, and metallic mercury. As a result, she obtained compounds such as nitrate and cobalt oxide, in addition to chloride, and described some of their properties. However, the term "cobalt" never appears in her books. For example, the chloride is mentioned as *Rosen-Farb Saltz* ("pink salt"). As a curiosity, it should be noted that "cobalt" is derived from the German word *Kobold* which means elf. In the silver mines of Saxony and Bavaria, from the fifteenth century onward, this term had come to be used to designate ores of low value. Miners believed that the latter ones were the result of malicious goblins acting on silver ores. (Martine de Bertereau also alludes to small creatures that inhabited the Earth's interior—see above.)

The thermochromism of cobalt compounds observed by Wallich is still an excellent teaching resource to introduce chemistry students to the principles of chemical equilibrium. Additionally, it has been used in some ambient temperature sensors, such as the well-known "weather roosters," which are blue on hot days and pink on cold days. However, for Wallich, such reversible color changes, induced by temperature variation, meant the *cauda pavonis* ("peacock's tail"), one of the stages in obtaining the Philosopher's Stone (the *Magnum opus*, or Great Work), characterized by the appearance of iridescence, which, it was believed, indicated that one was on the right path [55].

After the publication of her books, Wallich became a much sought-after alchemist, serving three German princes. In the contracts signed with them, she proposed to produce gold from silver. However, her incapacity to meet the contractual obligations not only ruined her reputation but also took away almost all of her assets. She lived the last 15 years of her life in the city of Arnstadt, in Thuringia, where, already a widow and childless, she died in poverty [56].

2.4   Heretics

As discussed before in this chapter, alchemy had been banned by a papal bull in 1317. However, despite Church prohibitions and condemnations, its practice consolidated in European courts throughout the fifteenth century [57]. Nevertheless, its association with demonic practices led to a further intensification of banishment following the reformation. To understand the growing demonization Europe went through, one must consider the magical atmosphere of the renaissance, largely influenced by the Italian Marsilio Ficino (1433–1499), a humanist philosopher in the service of Cosimo de Medici the

Elder, ruler of Florence. According to the intellectual climate of the time, Cosimo collected ancient Greek manuscripts, which Ficino translated. Around 1460, he stumbled upon a document found in a monastery in Macedonia, describing a magical-religious cult from Egypt. Believing in the existence of a lost wisdom of the ancients, the humanists considered these writings, dating back to around 1300 BC, to be attributed to the first Egyptian sage, Hermes Trismegistus (see Chap. 1), whom they compared to Moses. Ficino was convinced that he had found the true source of Western wisdom, the *prisca philosophia* (ancient philosophy). This knowledge, considered the first and purest of all, predated the Greeks and was believed to hold the key to understanding Nature and its hidden powers. For a century and a half, these Hermetic texts of Neoplatonic influence, collectively known as *Corpus Hermeticum*, formed the foundation for a natural philosophy imbued with magic, challenging Aristotelian thinking [58]. It embraced an organicist philosophy, viewing the world as a living organism with both a material body and an immaterial soul. The conviction that magi could control the soul of the world, by suppressing the influence of the stars, laid the basis for hermetic magic. What distinguished hermetic magic from astrology was that astrologers merely read the influences of the planets without attempting to alter them, whereas magi sought to actively manipulate and harness these celestial forces [59, 60].

As a cleric, Ficino was aware that defending the use of magic led him into theologically risky territory. He emphasized that it was a magic that only evoked natural forces and limited its application to issues of health. Among his most audacious followers was Giovanni Pico della Mirandola (1463–1494), an Italian who introduced Jewish Kabbalah into the Hermetic tradition. According to della Mirandola, magic, viewed as a sacred quest, expressed the magnanimity of God.

The culture of the Late Middle Ages had been marked by the harmonization of Aristotelianism with Christianity, and during the Renaissance, intellectuals like Ficino and della Mirandola had a similar desire regarding Hermeticism. However, despite initial acceptance, even within the Church, the Hermetic doctrine eventually faced opposition. Adding to the alleged interference with the effects of the stars, della Mirandola's allusion to angelic magic was seen as too close to the demonic. In 1487, many of his theses were condemned by a commission appointed by Pope Innocent VIII, and he was even arrested. The influential treatise *Malleus Maleficarum* [*The Hammer of the Witches*, 1486] by the Dominican inquisitor Heinrich Krämer dates from this time. This work became (in)famous for the methods of torture it suggests to

obtain confessions from practitioners of witchcraft, who were often sentenced to death.

The Church's hardened position after the Council of Trent in 1563 extended to popular magic as well. Between the late sixteenth and early seventeenth centuries, Europe experienced an intense wave of repression of both intellectual and popular magic, leading to a relentless hunt for its practitioners. This action by the clergy was part of a broader program to exterminate all forms of religious heterodoxy. With the reformation, in addition to the split between Protestants and Catholics, numerous sects had emerged (Anabaptists, Baptists, Puritans, Quakers, etc.), some of which gave women active roles in religious life, namely, preaching and the interpretation of God's Word. Both Catholics and Protestants, who considered the sects as incubators of blasphemous ideas, tended to intensify the association of religious heterodoxy with the female sex. Therefore, it was not uncommon for any religious fervor to generate suspicions of association with magic [61]. Women, especially poor women who behaved differently, spoke openly, or challenged male authority, were easily suspected of practicing witchcraft. The same suspicions arose if they were knowledgeable in herbal medicine, obstetrics, divination, incantations, or alchemy. Their alleged mental weakness and lasciviousness led people to believe that such knowledge could only be acquired through (sexual) pacts with Satan. While in theory men could also be accused of witchcraft, records show that the majority of victims were women, and in some villages, nearly the entire female population was sentenced to death [62, 63]. Popes, Protestant reformers, Counter-Reformation agents, humanists, and magistrates created a paranoia among the population surrounding demonic practices. Exact figures are difficult to determine, but it is estimated that in the sixteenth and seventeenth centuries alone, around sixty thousand people, mostly women, were sentenced to death on charges of witchcraft [64, 65].

The engraving of Hans Baldung Grien shown in Fig. 2.9—three witches brewing potions during a sabbath, while another one flies on a goat—is representative of the vast iconography of the sixteenth and seventeenth centuries that tended to portray women as lewd, insolent, and troublemakers, who caused disorder [66].

One of the most famous cases of the condemnation of an alchemist to death by burning was that of Anna Maria Zieglerin (1550–1575), in sixteenth-century Germany. Married to Heinrich Schombach, assistant to Philipp Sömmering, an alchemist in the service of the Prince of Braunschweig-Wolfenbüttel, Anna Zieglerin had her own laboratory. In 1573, she presented to the prince a manuscript of her own with 20 pages and the title "On the noble and valuable art of alchemy." In it, she taught how to obtain the

Fig. 2.9 The Witches (1510), Hans Baldung Grien. Metropolitan Museum of Art, New York

Philosopher's Stone, as well as how to imitate precious stones and how to prepare medicines. However, in 1575, the impossibility of meeting the expectations of the employer, desirous of gold, led to their execution—Anna Zieglerin for witchcraft and adultery, and the men for fraud and treason. One of the points of the incrimination concerned the fabrication of a story about a supposed adept of alchemy, whose secrets would enrich the prince's court.

She claimed that such a character, her mentor, a count named Carl von Oettingen, allegedly the illegitimate son of Paracelsus, had transmitted many of the alchemical secrets inherited from his father to her. After several torture sessions, the trio eventually admitted to the crimes they were accused of. Before being quartered alive, the two men were pinched with red-hot tongs. Anna Zieglerin, who received the same treatment, was subsequently tied to an iron bench and burned alive [67, 68].

References and Notes

1. Christine de Pisan, *La Cité des Dames,* Stock, Paris, 1986, p. 90
2. U. Eco, "Introdução à Idade Média", *in* Umberto Eco (org.) *Idade Média—Explorações Comércio e Utopias,* Vol. IV, Dom Quixote, Lisboa, 2011, p. 15
3. Ruth Watts, *Women in Science—A Social and Cultural History,* Routledge, Abingdon, 2007, p. 20
4. The introduction of monasticism in Europe, due to Athanasius of Alexandria (*c.*296–373), one of the Church Fathers, took place around the year 340. However, it is Benedict of Nursia (*c.*480–550), founder of the Benedictine Order, who is usually considered the father of Western monasticism, due to the strong influence he fad in its spread. In turn, the origin of female monasteries dates back to the time of Saint Macrina (*c.*330–379), a nun who created in Pontus (a territory in present-day northeastern Turkey, on the southern coast of the Black Sea) one of the first monastic communities in Christendom. But it was Scholastica, sister of Benedict of Nursia, who founded, in Piumarola, Italy, the first Benedictine Monastery for women, whose example was to be followed throughout Europe. *Vide* P. A. Tamanini, "The Ancient History and the Female Christian Monasticism: Fundamentals and Perspectives", *Athens Journal of History,* 3 (2017) 235–250
5. S. Kersey, "Medieval education of girls and women", *Educational Horizons,* 58 (1980) 188–192
6. Leigh Ann Whaley, *Women's History as Scientists—A Guide to the Debates,* ABC, CLIO, Santa Barbara (CA), 2003, p. 33
7. Political union of several territories in Western, Central and Southern Europe, created in the Early Middle Ages and which dissolved in 1806 with the Napoleonic Wars.
8. Wighard Strelow, Gottfried Hertzka, *Hildegard of Bingen's Medicine,* Bear & Company, Rochester, VT, 1988, p. xi
9. C. Singer, "The visions of Hildegard of Bingen", *Yale Journal of Biology and Medicine,* 78 (2005) 57–82
10. M. R. N. Costa, "Mulheres Intelectuais na Idade Média: Hildegarda de Bingen—Entre a Medicina, a Filosofia e a Mística", Trans/Form/Ação, 35 (2012) 187–208

11. Margaret Alic, *Hypathia's Heritage—A History of Women in Science from Antiquity to the Late Nineteenth Century*, The Women's Press, London, 1990, pp. 62–76

12. M. H. Green, "In Search of an «Authentic» Women's Medicine: The Strange Fates of Trota of Salerno and Hildegard of Bingen", *Dynamis. Acta Hisp. Med. Sc. Hist. Illus.*, 19 (1999) 25–54

13. Merry E. Wiesner-Hanks, *Women and Gender in Early Modern Europe*, Cambridge University Press, Cambridge, 2000, p. 216

14. Whaley, *Op. cit.*, (6) pp. 34–36

15. M. H. Green, "In Search of an «Authentic» Women's Medicine: The Strange Fates of Trota of Salerno and Hildegard of Bingen", *Dynamis. Acta Hisp. Med. Sc. Hist. Illus.*, 19 (1999) 25–54

16. P. Cavallo, M. C. Proto, C. Patruno, A. Del Sorbo, M. Bifulco, "The first cosmetic treatise of history. A female point of view", *International Journal of Cosmetic Science*, 30 (2008) 70–86.

17. Alic, *Op. cit.* (11), pp. 50–58

18. J. W. Scott, ""La querelle des femmes" no final do século XX", *Estudos Feministas*, 9 (2001) 367–388

19. J. Kelly, "Early Feminist Theory and the "Querelle des Femmes", 1400–1789", *Signs*, 8 (1982) 4–28

20. Whaley, *Op. cit.* (6), p. 44

21. M. Karawejczyk, "Christine de Pisan, uma feminista no medievo?!", *Históriæ*, 8 (2017) 189–203

22. Whaley, *Op. cit.* (6), p. 45

23. Phyllis Stock, *Better Than Rubies—A History of Women's Education*, Putnam, New York, 1978, pp. 40–47

24. Whaley, *Op. cit.* (6), p. 41–47

25. Anatole France, *Rabelais*, Calmann-Lévy, Paris, 1928, p. 79

26. F. Rigolot, "Écrire au féminin à la Renaissance: Problèmes et perspectives, *L'Esprit Créateur*, 30 (1990) 3–10

27. M. K. Papathanassiou, "The Occult Sciences in Byzantium", *in* Stavros Lazaris (ed.), *A Companion to Byzantine Science*, Brill, Leiden (2020) pp. 464–495

28. Alic, *Op. cit.* (11), p. 48

29. António Amorim da Costa, *Ciência e Mito,* Imprensa da Universidade de Coimbra, Coimbra, 2009, p. 9

30. K. Park & L. Daston (eds.), *The Cambridge History of Science*, Cambridge University Press, Cambridge, 2006, vol. 3, pp. 497–498

31. G. Devincenzo, "Dans les coulisses de l'atelier d'un maître verrier, ou Marie de Gournay et les séductions de la science", *Studi Francesi*, 179 (2016) 193–201

32. Allen G. Debus, *The Chemical Philosophy: Paracelsian Science and Medicine in the Sixteenth and Seventeenth Centuries*, Dover Publications, Newton Abbot, 2002, p. xxi

33. J. R. Partington, *A Short History of Chemistry*, Dover Publications, New York, 1989, p. 42

34. W. R. Newman, L. M. Principe, "Alchemy *vs.* chemistry: The etymological origins of a historiographic mistake, *Early Science and Medicine*, 3 (1998) 32–65

35. M. E. Warlick, "Moon Sisters: Women and Alchemical Imagery", *in The Golden Egg: Alchemy in Art and Literature* (Leipzig Explorations in Literature and Culture), Alexandra Lembert, Elmar Schenkel (eds.), Galda und Wilch, 2002, pp. 183–197

36. Jan Apotheker, Livia Simon Sarkadi (eds.), *European Women in Chemistry*, Wiley, Weinheim, 2011, pp. 9–11

37. João Paulo André, *Poções e Paixões—Química e Ópera*, Gradiva, Lisboa, 2018, p. 178

38. Jean-Pierre Poirier, *Histoire des Femmes de Science en France*, Pygmalion, Paris, 2002, p. 101

39. Martine de Bertereau, *La restitution de Pluton*, Hervé du Mesnil, Paris, 1640, p. 7

40. I. M. P. Valderrama, J. Pérez-Pariente, "Alchemy at the service of mining technology in seventeenth-century Europe, according to the works of Martine de Bertereau and Jean du Chastelet", *Bull. Hist. Chem.*, 37 (2012) 1–13

41. G. Kass-Simon, Patricia Farnes (eds), *Women of Science*, Indiana University Press, p. 305

42. Claude Viel, "Le Dictionnaire de chimie de Pierre-Joseph Macquer, premier en date des dictionnaires de chimie. Importance et édtions successives", Rev. Hist. Pharm., LII, 342 (2004) 261–276

43. Poirier, *Op. cit.* (38), p. 93

44. Along with the Philosopher's Stone and the Elixir of Life, the homunculus (a small human being created in the laboratory) was the third great design of the alchemists.

45. Poirier, *Op. cit.* (38), p. 94

46. Quoted by Poirier, *Op. cit.* (38), p. 84

47. M. de Gournay, *Les Advis, ou les Présents de la Damoiselle de Gournay*, Paris, 1641 (3ª edição), p. 929

48. M. Deslauriers, "One soul in two bodies—Marie de Gournay and Montaigne", *Journal of Theoretical Humanities* (2008) 13, 5–15.

49. P. Bayer, "Women Alchemists and the Paracelsian Context in France and England, 1560–1616", *Early Modern Women*, 15 (2021) 103–112

50. P. Bayer, "Madame de la Martinville, Quercintal's Daughter and the Philosophers Stone: Manuscript Representations of Women Alchemists", *in* Kathleen Perry Long (ed.), *Gender and Scientific Discourse in Early Modern Culture*, Ashgate, Farnham 2010, cap. 7

51. A. Kraft, "Dorothea Juliana Wallich (1675–1725) and Her Contributions to the Chymical Knowledge about Element Cobalt", *in* Annete Lykknes, Brigitte Van Tiggelen (eds), *Women in their Element—Selected Women's Contributions to the Periodic System*, World Scientific, Singapore, 2019, p. 59

52. *Ibid.*, pp. 61–64

53. *Schlüssel zu dem Cabinet der geheimen Schatz-Cammer der Natur, zur Such- und Findung des Steins der Weisen, durch Fragen um Antwort gestellet*

54. Solid cobalt(II) chloride at room temperature is hexahydrate ($CoCl_2.6H_2O$), and it is pink. Upon heating, the water molecules are released into the atmosphere, originating the anhydrous form, $CoCl_2$, which is blue. In aqueous solution the thermochromism is different (although the color change is similar), as the pink and blue forms are due, respectively, to the complex ions $[Co(H_2O)_6]^{2+}$ and $[CoCl_4]^{2-}$ which are in equilibrium. This equilibrium is temperature dependent: an increase in temperature favors the formation of the blue complex; a decrease in temperature favors the formation of the pink complex. The intermediate colors obtained by Wallich were due to the presence of traces of nickel and iron in the samples used.

55. Originally, the process of obtaining the Philosopher's Stone included four stages: black, white, yellow and red, which date back to the first centuries of the Christian era (see Chap. 1). Later on, other stages would be mentioned, namely the *cauda pavonis* (peacock's tail), with a variety of colors.

56. Kraft, *Op. cit.* (51), p. 64–66

57. Antonio Clericuzio, "Alquimia e Química Empírica", *in* Umberto Eco (org.) *Idade Média—Explorações Comércio e Utopias*, Vol. IV, Don Quixote, Lisboa, 2011

58. In 1614, a philological investigation of the *Corpus Hermeticum* showed that it did not date back to ancient Egypt, but rather from the last years of the Roman Empire.

59. James Hannam, *A Origem da Ciência—Como os Filósofos do Mundo Medieval Lançaram os Fundamentos da Ciência Moderna*, Alma Livros, Loures, 2021, pp. 245–248

60. Margaret Wertheim, *Pythagoras's Trousers—God, Physics, and the Gender Wars*, Fourth Estate, London, 1997, pp. 83–85

61. *Ibid.*, pp. 86–90

62. Carolyn Merchant, *The Death of Nature—Women, Ecology and the Scientific Revolution*, Harper & Row, New York, 1983, p. 138

63. L. Tosi, "Mulher e Ciência—A Revolução Científica, a Caça às Bruxas e a Ciência Moderna", *Cadernos Pagu*, 10 (1998) 369–397

64. M. D. Bailey, "From Sorcery to Witchcraft: Clerical Conceptions of Magic in the Later Middle Ages", *Speculum*, 76 (2001) 960–990

65. R. A. Horsley, "Who Were the Witches? The Social Roles of the Accused in the European Witch Trials", *The Journal of Interdisciplinary History*, 9 (1979) 689–715

66. Merchant, *Op. cit.* (62), pp. 134–138

67. T. E. Nummedal, "Alchemical Reproduction and the Career of Anna Maria Zieglerin", *Ambix*, 48 (2001) 56–68

68. Diana Maury Robin, Anne R. Larsen, Carole Levin (eds.), *Encyclopedia of Women in the Renaissance: Italy, France, and England*, ABC-CLIO, Santa Barbara (CA), 2007, pp. 5–6

3

Chastes and Keepers of Secrets

"O God", said Leonora, "you have not yet found the remedy to which I refer. You find so many remedies for bad blood and anger, yet you never purge the stomachs and blood of men, who are always sick of heart and brain. If only the remedy could be found to cure us of the simplicity, piety, and love we undeservedly bring to these sick ones of ours".

"Galen did not describe such a medicine, nor has any other author ever found it", said Corinna. "Or, if he did, he didn't leave it in writing, which was not his business, because wolves do not eat wolves. They know too well what their damage would be if they hadn't us".

MODERATA FONTE, *Il Merito delle Donne* (1600) [1].

Sixteenth-century Italy was rich in books of "secrets." Among these works, Giambattista della Porta's *Magia Naturalis* stands out as one of the best models. These compilations consisted of recipes, formulations, and practical advice covering various subjects like natural philosophy, astrology, medicine, alchemy, cosmetics, cooking, and occultism. Dating back to the period before the printing press, the manuscript *Experimenti* is another remarkable work, revealing the great interest of Renaissance courts in collecting recipes and practical secrets. Its authorship is attributed to Caterina Sforza, the founder of the Grand Ducal dynasty of the Medici.

The feminine counterpart to printed books of secrets was inaugurated in Venice in 1561 with *I Secreti della Signora Isabella Cortese*. This work comprises more than four hundred recipes spanning fields such as medicine, alchemy, cosmetics, and preparations for domestic use. In England, the first female authorship of these books emerged in the mid-seventeenth century,

J. P. André, *Sisters of Prometheus*, https://doi.org/10.1007/978-3-031-57136-7_3

influenced by the *London Pharmacopoeia* published in 1618. Alethea Talbot's *Natura Exenterata* was distinctive among the others due to its stronger emphasis on scientific content.

3.1   *Experimenti*

With the collapse of the feudal regime, the fifteenth and sixteenth centuries witnessed significant urban, social, and economic changes. The Renaissance, which originated in fourteenth-century Italy with Florence as its epicenter, sparked a renewed interest in Greco-Roman antiquity. This gradually diminished the Church's dogmatic influence on society, fostering a deeper appreciation of humanity and the natural world. In his work *De Hominis Dignitate Oratio* [*Oration on the Dignity of Man*, 1486], Pico della Mirandola (see Chap. 2) emphasized the greatness of the human being as an intermediate reality between the world and God, making it a true manifesto of the Renaissance spirit.

The idea of perfection bestowed upon man, the greatest divine creation, was not extended to women, who continued to be considered inferior in practice. Despite the Renaissance's intense pursuit of knowledge, only aristocratic women were deemed to require a thorough education, leading to no noticeable changes in the education of ordinary women. At most, the latter were sometimes allowed to learn only what was strictly necessary to become good Christians, exemplary wives, and mothers. This reflected some influence from the ancient Roman model, which assigned mothers the duty of preparing their male children for citizenship (see Chap. 1). The emphasis on motherhood and the reproductive role of women, characteristic of the Renaissance, to some extent, reflected the consequences of the Black Death, which, between 1347 and 1353, may have decimated at least one-third of the European population.

The ideal of the educated woman was defined by humanist treatises such as *De Studiis et Litteris* [*On Studies and Litera*ture], which originates from the first half of the fifteenth century and stands as the first work entirely devoted to female education. In this treatise, the author, Leonardo Bruni, establishes a connection between intellectual education and the formation of character, similar to what would be found in comparable works in the following century. While considering Latin to be the basis of any study, and believing that knowledge was important only to women of the nobility, Bruni recommended that they read the Church Fathers, as well as Virgil, Cicero, and Livy. In contrast, Vittorino da Feltre, the most famous Italian pedagogue of the fifteenth century, outlined what they should learn: Latin and Greek, fundamental tools for understanding literature, philosophy, and the history of antiquity, along with declamation, arithmetic, geometry, astronomy, and logic. However, he placed a significant emphasis on Christian humanism. In the next century, *Il Libro del Cortegiano* [*The Book of the Courtier*, 1528], by Baldassare Castiglione (who dedicated it to the Portuguese Bishop D. Miguel da Silva) [2], has been the work that most influenced the education of aristocrats at the beginning of European modernity. This narrative about the court of Urbino, in the form of a philosophical dialogue, became a manual of court etiquette and ethics. For Castiglione, women of the nobility should possess the same qualities as the men—prudence, grandeur, thoughtfulness, kindness, discretion—as well as the ability to educate their children and manage their husband's estates. They should be literate, have knowledge of music, and know how to draw and paint. Their education, in addition to molding their character, would allow them to have pleasant conversations. This, in turn, would serve as an adornment to their husband's court, thus complementing his role as a courtier. However, chastity, graciousness, and modesty were virtues much more appreciated in women than their knowledge or their intellectual attributes [3, 4].

Caterina Sforza (1463–1509) was educated in all of that, but she achieved much more (Fig. 3.1). She was the daughter of Galeazzo Maria Sforza, Duke of Milan, and his lover Lucrezia Landriani, having grown up at her father's court. From an early age, she distinguished herself with beauty, intelligence, determination, and bravery—qualities befitting the granddaughter of Francis Sforza, the famous *condottiero* and creator of the powerful Milanese dynasty [5]. Caterina Sforza went on to become the founder of the Grand Ducal dynasty of the Medici, immortalized by Niccolò Machiavelli in *The Prince* (1513) and *Discourses* (*c.*1517). Her life and achievements make her one of the most remarkable and fascinating women of the Italian Renaissance [6].

At the age of 14, she married Girolamo Riario, the favorite nephew of Pope Sixtus IV. In 1484, while pregnant with her fourth child, upon learning of the death of the supreme pontiff, she promptly assaulted Castel Sant'Angelo in Rome, intending to exert pressure on the papal conclave [7]. The plan failed, and the couple ended up retiring to Forlì, a dominion in Emilia-Romagna over which Girolamo ruled as Lord. However, once there, events were not favorable to them either: in 1488, Girolamo was brutally murdered by political enemies, and Caterina was imprisoned with her children. She managed to escape, leaving the children in the hands of the adversaries who immediately threatened to take their lives. According to Machiavelli, as she lifted her dress and showed her genitals, she retorted that if the children were killed, she had what she needed there to generate others [8]. With the help of the military forces of her uncle Ludovico Sforza, she managed to control the rebellion and henceforth govern Forlì as regent. She is said to have married Giacomo Feo, a

Fig. 3.1 *Caterina Sforza* (c.1481), Lorenzo di Credi. Pinacoteca Civica de Forli

loyal servant of her late husband, in secrecy. However, Giacomo was also murdered in 1495. Persistent and ambitious, she had a third relationship (and perhaps marriage), this time with Giovanni de' Medici, known as il Popolano, with whom she had a son, Giovanni delle Bande Nere [9]. When the Venetian forces invaded their territories, it was herself who trained her own troops, a fact that earned her the nickname "Tigress." The next invader, in 1499, was Cesare Borgia, son of Pope Alexander VI. After intense fighting, she was taken prisoner in January 1500 and eventually released in June of the following year, on the condition that she renounced her claims to Imola and Forlì. She spent her last years in Florence, where she devoted herself to family and alchemy. She died in 1509 at the age of 46, probably a victim of malaria, and was buried in Le Murate, a monastery where she had always kept a cell for spiritual retreats [10, 11].

Caterina Sforza was the author of a famous manuscript known as *Experimenti* [12]. It reveals how the (al)chemical procedures applied to health and physical appearance were activities in which several women from the beginning of modernity were actively involved [13]. It consists of 454 recipes, mostly for medicines (for conditions ranging from fever to malignant tumors) and perfumery and cosmetic preparations. For example, it includes remedies to lighten hair, according to the Renaissance ideal of female beauty that Agnolo Firenzuola extoled in the *Dialogo delle Bellezze delle Donne* [*Dialogue on the Beauty of Women*, 1541]. It should be noted that as early as the first century BC, Ovid, one of the greatest exponents of Roman literature, in his didactic poem *Medicamina Faciei Femineae* [*Cosmetics for the Face of Women*], had already advocated the use of these products.

Despite what its title might suggest, Caterina Sforza's *Experimenti* is nothing more than a collection of practical procedures and cannot be equated with experimentation in the modern sense. In the sixteenth century, the Latin terms *experimentum* and *experientia* were practically synonymous; their distinction would only be properly established in the course of the eighteenth century [14]. In the medieval sense, *experimentum* could either refer to any practical knowledge acquired directly from experience [13] or to a change in the course of an event by the action of a procedure or the application of previously tried formulas, as well as the rational use of observation to resolve questions of a philosophical nature [15]. From Galileo (1564–1642) on, *experimentum* also acquired the modern meaning of experimentation, namely, a controlled observation intended to test a hypothesis [16].

Besides reflecting the influence of oral tradition, the recipes of *Experimenti* also reveal the author's diverse sources of knowledge, including her readings in alchemy, astrology, and natural philosophy, as well as her own experience, which were enriched through interactions with numerous correspondents. Notably, the letters she exchanged with the *speziale* (apothecary) of Forlì, Lodovico Albertini, illustrate her avid pursuits of secrets and formulations. Furthermore, *Experimenti* demonstrates the author's keen interest in applying chemical procedures to pharmacy and cosmetics and the attraction for methods, techniques, and tricks that could bolster her political power. She sought to achieve this through the production of alchemical gold and counterfeit coins or by preparing poisons and their corresponding antidotes. In fact, Caterina Sforza faced accusations of sending a poisoned letter to Pope Alexander VI in 1499 [17]. While most of the recipes of her *Experimenti* are written in Italian, some are in Latin, particularly those related to the transmutation of metals. It is possible to imagine that the use of Latin was intended to underscore the seriousness and significance of these recipes. Conversely, certain alchemical formulations, as well as recipes of a sexual nature (such as those addressing impotence or lack of libido), are encrypted, with the decryption key, however, provided in the manuscript [18].

The ingredients for the medicinal recipes of *Experimenti* are primarily of botanical origin. It is possible that Caterina Sforza was introduced to the world of pharmaceutical plants through her stepmother, Bona Maria de Savoia, to whom she was very attached, as well as through her *speziale*, Cristoforo de Brugora. The translation from Greek into Latin made by Ermolao Barbaro at the end of the fifteenth century of *De Materia Medica*, authored by the Greek botanist Pedanius Dioscorides in the first century, was followed in the sixteenth century by an annotated version by the Italian physician Andrea Mattioli. This resurgence of medicine based on plant products also played a significant role in the establishment of botanical gardens [19]. In this context, a grandson of the "Tigress of Forlì," who designed her own gardens, would have a crucial role (see below). The small jasmine vase depicted in her magnificent portrait by Lorenzo di Credi (Fig. 3.1) may symbolize her skillful use of medicinal plants. Furthermore, concerning her knowledge of their properties, it is worth noting that she maintained a connection with the Florentine Monastery of Le Murate until the end of her days, as evidenced by the correspondence exchanged with Abbess Elena Bini. Typical of monastic institutions, this monastery had a dedicated facility for the preparation of therapeutic products [20].

Here are some examples of recipes found in *Experimenti*, beginning with those crafted for medicinal purposes, such as pills, powders, ointments, distillates, elixirs, antidotes, and more, which hold a dominant presence in the manuscript. One such case is the *acqua mirabile et divina* (prodigious and divine water), known for its supposed effectiveness against toothaches, earaches, melancholy, lack of memory, as well as conditions like leprosy and other diseases. This particular recipe underscores the significance of color changes in alchemy, as discussed in Chap. 1. To create it, a combination of various botanical varieties, spices, and brandy is distilled over several days, resulting in three distinct fractions, each with progressively enhanced therapeutic properties. The third fraction, distinguished by its coveted red color, is considered the most virtuous liquid [21]. Another *acqua mirabile*, capable of combating infections, promoting liver function, and treating malignant tumors and paralysis, also relies on plant-based ingredients such as nutmeg, cardamom, galangal, long pepper, and white wine, supplemented with the addition of antimony (a metal commonly used in alchemy and medicine as an emetic). The resulting mixture is then pulverized and distilled using a gentle heat in a still [22].

Experimenti also provides recipes and formulas for addressing issues related to sexuality and reproductive function. However, these are not included in the printed edition of the manuscript, which was produced in 1893 through the initiative of the politician, historian, and writer Pier Desiderio Pasolini. For instance, to restore a woman's virginity, *Experimenti* suggests distilling sage in water and applying the resulting liquid as needed to the desired area of the body. Despite the simplicity of this recipe, it is described as *verissima e maraviglia* (truly wonderful). Additionally, to enhance male sexual performance, *Experimenti* offers recipes based on ingredients considered aphrodisiacs. These recipes include products meant to be ingested with wine as well as preparations for direct application to the sexual organ. As an example, one such recipe calls for orchid root, wild boar fat, pepper, and skink (a type of lizard that, reduced to powder, was in great demand at the time) [18].

The 66 recipes for cosmetic preparations presented in *Experimenti* reveal the high interest of Renaissance aristocrats in beauty secrets. As Castiglione observes in his *Il Libro del Cortigiano*, "no court, however great, can have ornament, splendor, or joy, without ladies" [23], to which he adds that "a woman without beauty is lacking much" [24]. Thus, using recipes such as those of Caterina Sforza, it was possible to compose a profusion of waters and skin lotions, as well as products to color the hair, lips, and face. These formulations, like those for therapeutic purposes, mainly required ingredients of vegetable origin but also included components from the mineral kingdom, such as metals and precious stones. One notable example is the recipe of *acqua de talcho* (talc water), which involved the distillation of calcined talc (hydrated magnesium silicate). Allegedly, this preparation could make a 60-year-old woman appear 20. Reduced to powder and dissolved in white wine, this *acqua* acted as an antidote to plague and poisons, while the water itself could make pearls of inferior quality shine brighter and even increase their size [25]. In turn, the recipe for "water to protect the skin against stains" (*un'acqua mirabili aconservare il viso contra omni macula*) required the distillation of fennel, betony, endive, roses, and white wine. According to the alchemical belief that the longer the distillation, the better the properties of the product obtained, over the course of the days one would sequentially obtain pink water, silver water, and, finally, the "extraordinary golden water" [26].

Several of the recipes in *Experimenti* require actual chemical substances. For example, there is a preparation to "whiten the face and give it beauty and luminosity" (*a fare la faccia bianchissima et bella et lucente*), which is based on silver litharge (lead(II) oxide with traces of silver). This substance should be pulverized and dissolved in vinegar and then distilled [27]. Another "beauty water" (*acqua perfettissima per far bella*) involves the distillation of a mixture of apples, turpentine, white wine, and mastic. In addition to gum tragacanth and goat's milk, the recipe also calls for mercury and sage to be pulverized in a stone mortar with a walnut pestle [28].

The 30 recipes of strict alchemical scope in the manuscript include procedures to give the appearance of gold to base metals or to increase the weight of current coins, thereby increasing their value [29]. In 1504, an unidentified correspondent of Caterina Sforza who referred to himself as "Your faithful servant" requested from her the recipe for transmuting base metals into 19-carat gold in exchange for a face-whitening preparation [30].

Caterina Sforza bequeathed the manuscript of *Experimenti* to her youngest son, Giovanni delle Bande Nere, who was not only a famous *condottiero* but also the father of Cosimo I de' Medici, Grand Duke of Tuscany. It can thus be said that the origins of the long-standing connection between the Medici family and alchemy and natural philosophy trace back to her. This connection first became evident with her grandson's establishment of the foundry in the Palazzo Vecchio in Florence. Later, her great-grandson, Francesco I, installed a laboratory in the Casino Mediceo di San Marco in the same city. In this laboratory, Francesco I, accompanied by his wife, Bianca Cappello, spent extended periods (see Chap. 2) conducting experiments with materials such as porcelain, enamel, and faience. They distilled medicinal waters, produced incendiaries (early forms of matches), and even created counterfeit jewelry [31]. After the death of his son, Antonio de' Medici, inventories revealed the presence of alchemical treatises, various collections of recipes, as well as equipment and materials for experimentation [32]. It was also Cosimo I who, in 1544, established the first botanical garden in Europe, in Pisa. The following year, he founded the Giardino dei Semplici in Florence, which is now part of the Museum of Natural History of the University of Florence. Both gardens were designed by the physician and botanist Luca Ghini, originally from the Imola region. In turn, the Boboli Gardens (Fig. 3.2), constructed in Florence in 1550 by Leonor de Toledo, the wife of Cosimo I, were a testament to the mastery of Nature through science and art. It is worth noting that this garden was built in an area without natural water sources [33].

Currently, only one handwritten copy of *Experimenti* is known and it belongs to a private archive in Ravenna. It is a transcription from the original, made by Lucantonio Cuppano, a soldier in the service of Giovanni delle Bande Nere. With a total of 553 pages and bound in leather, it contains the following annotation: "This parchment was obtained by Gio[vanni] de Medici ... inherited from his mother Caterina Sforza Riario, 1514." Cuppano's copy was published in the nineteenth century by Pier Desiderio Pasolini, biographer of Caterina Sforza, who considered it the "most complete and important document known on perfumery and medicine from the beginning of the

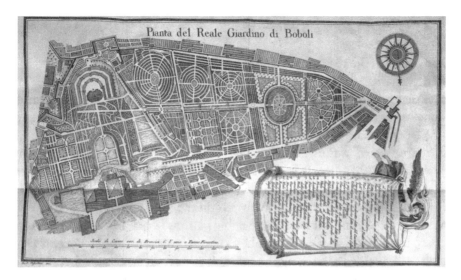

Fig. 3.2   Boboli Gardens, Florence. Engraving from the eighteenth century

sixteenth century" [34]. Given the nonexistence of the original manuscript, we must rely on the testimony of Cuppano, both in his preface and in various parts of the text in which he reiterates the aristocrat's authorship. In the preface, he also emphasizes that the recipes were tried by herself. The letters she exchanged with the *speziale* of Forlì (to whom she left a debt of at least 587 florins), as well as with other interlocutors, provide further evidence of *Experimenti* being her own work [35]. The recent discovery at the Biblioteca Nacional Florence Central of an unpublished manuscript with more than four hundred of recipes of hers, also copied by Cuppano, as well as an index that may have belonged to another collection of recipes of hers, sheds new light on the experimentalist activity of the "Tigress of Forlì" (including in the field of veterinary medicine), contributing to a better understanding of domestic medicine in the courts of the Italian Renaissance [36].

The Italian archives are, indeed, rich in manuscript collections of recipes, reflecting the avidity of that time for the collection of practical (al)chemical secrets, applied to everyday life. Another work that, along with *Experimenti*, circulated among the educated female audience in the beginning of the sixteenth century was *Ricettario Galante* [*Galante Recipe Book*], whose subtitle is *Book to become beautiful* [37].

3.2 *Segreti*

The strong desire to acquire valuable *secrets* led to active networks of information exchange throughout the sixteenth century. This exchange of *secrets* occurred not only through oral communication but also in written form. In some cases, these closely guarded *secrets* were compiled into collections, such as *Experimenti* by Caterina Sforza. Additionally, it is important to note that the term "secret" was used interchangeably with "experience," not in the sense of verifying a hypothesis but rather as a previously tried formula, as mentioned above [16]. While the compilation of recipes was not exclusive to palatial environments, it thrived notably within the royal and aristocratic settings of kings, dukes, and princes. These noble patrons frequently supported alchemists, magi, and natural philosophers, hoping to gain material benefits from their well-guarded secrets and knowledge. An excellent historical example of such patronage is seen in the support provided by Rudolph II, who served as the Holy Roman Emperor from 1575 to 1612 [31].

Initially, collections of practical secrets were handwritten, resulting in limited circulation. However, the advent of the printing press brought them widespread popularity, a popularity that endured until the eighteenth century (when the appearance of periodical scientific journals led to their decline). These collections had ancient roots, tracing back to manuscripts copied and translated into Latin in medieval monasteries. However, while for the medieval mindset all that could be known had already been discussed by classical authors, according to the Renaissance perspective, the scholastic tradition was exhausted, and there was original wisdom based on revelation. The prevailing idea was that Nature held hidden (invisible) forces that humans could manipulate if they possessed the appropriate formulas and techniques. It was this spirit that contributed to the success of the books of secrets, which promised to unlock the secrets of Nature. The latter would only remain hidden until revealed through experimentation, and once explored, they would offer material advantages and improve people's lives. Such books were strictly utilitarian and practical, focusing on teaching how to accomplish specific tasks rather than explaining principles. This Renaissance pragmatism gained momentum throughout the sixteenth and seventeenth centuries [38, 39]. Notable examples of these books include *Secreti* (1555) by Alessio Piemontese and *Magia Naturalis* (1558) by Giambattista della Porta. The latter, initially published in Latin, achieved enormous success, with at least five editions in the first 10 years and translations into several languages—Italian (1560), French (1565), Dutch (1566), and English (1658) [40]. In 1589, an expanded edition was released, featuring 20 sections covering various fields such as natural philosophy, medicine, home management, alchemy, magnetism, statics, cosmetics, perfumery, distillation, and pyrotechnics. This book, based on renowned authors of Antiquity such as Pliny and Theophrastus, reflects the scientific and technical knowledge that prevailed until the time when figures like Galileo Galilei emerged on the scene.

The title *Magia Naturalis* translates to *Natural Magic* in English. In the context of the Renaissance mentality, magic was not immediately associated with the supernatural as it is for us in the twenty-first century. Instead, it referred to the concept of "natural magic," which was believed to be responsible for many of the natural processes. This type of magic was based on the assumption that certain bodies possessed natural powers or elemental forces, though invisible, that could influence other bodies and give rise to phenomena inexplicable in terms of Aristotelian understanding. As a result, one can refer to this as "Renaissance naturalism," in contrast to other types of magic, particularly the demonic variety. This revitalization of the magical tradition took place between the late fifteenth century and the early seventeenth century, primarily due to the rediscovery of ancient texts authored by hermetic and neoplatonic scholars, as detailed in the preceding chapter [41].

As noted by William Eamon in his significant work, *Science and the Secrets of Nature: Books of Secrets in Medieval and Early Modern Culture*, natural magic thrived in numerous European courts and was considered the court science par excellence. According to this historian, the essential characteristics of the magus, including a passion for seeking secrets, a taste for rarities, a fascination with marvels, and an approach to scientific knowledge as a theatrical spectacle meant to delight and surprise spectators, aligned perfectly with the courtier model advocated by Castiglione in his idealized account of the court of Urbino [42]. The oldest courtly academies, which aimed to promote natural knowledge, were established with a particular interest in natural magic. Among them, the Accademia dei Lincei (Academy of the Lynxes), based in Rome and still in existence today, stands out. Its founder, Federico Cesi (1585–1630), drew inspiration directly from what Giambattista della Porta had written in the introduction of his *Magia Naturalis*, emphasizing the need to observe Nature with the keen eyes of a lynx to harness its secrets [43].

Interestingly, della Porta did not overlook his female audience and included multiple recipes for hair and eyebrow dyeing, concealing paleness, and eliminating wrinkles and warts in the ninth section of *Magia Naturalis*, which was dedicated to embellishment. Among the many secrets, one particularly intriguing was the ability to "make a deflowered woman a virgin again." To achieve this, the method involved using tablets containing mastic, arsenic(III) sulfide, and sulfuric acid, which, when applied, would cause ulceration as well as the desired bleeding [44]. It is noteworthy that della Porta faced challenges with the Inquisition, having to respond before the Tribunal of the Holy Office on two occasions (in 1574 and 1580) due to suspicions of practicing demonic magic. Consequently, he had to persistently fight for the publication of his works [45].

By the mid-sixteenth century, Venice had become a major publishing center for books of secrets in the Italian language. This prominence could be attributed to its high number of pharmacies and its role as an important entry point for exotic products into Europe. It was precisely in the Serenissima City that the first work of this genre written by a woman was published in 1561, under the title of *I Secreti della Signora Isabella Cortese* [*The Secrets of Signora Isabella Cortese*] [46]. According to the information provided by the author herself, the book (Fig. 3.3) is a compilation of recipes of the "hidden secrets of Nature," encompassing fields ranging from medicine and cosmetics to the domestic sphere and alchemy. Out of a total of 404 recipes, with a strong alchemical character, more than half are intended for health and beauty [47]. The book's success was remarkable: by 1677, there were already 11 Italian editions and three in German by the end of the seventeenth century [48].

Little or nothing is known about the author. Some have speculated that the name "Cortese" could be an anagram of *secreto* [49], while others suspect that a man may have hidden behind a female name [50]. From that time, there is a known case of Giovanni Marinello, the author of *Gli Ornamenti delle Donne* [*Ornaments of Women*, 1562], who, in order to capture the public's attention, declared in the dedication that he was not the actual author of the book. Instead, he claimed that its origin was based on the writings of an ancient Greek queen whose name had been lost to time [51]. However, the existence of the request for authorization for the publication (*previlegio a stampa*) of *I Secreti della Signora Isabella Cortese* in the Venice archives, to which the Senate agreed on August 17, 1560, reduces the uncertainties. Historian Claire Lesage points out that, in the second half of the sixteenth century, the region of Venice was an important center of women's literature, reinforcing the possibility that *I Secreti* may have indeed been written by a female hand [50]. Furthermore, it is not unlikely that a woman would have used writing as a means to disseminate her knowledge among her contemporaries, especially in the specific case of cosmetics. The sections devoted to medicine and alchemy may also have the same type of authorship, similar to what Caterina Sforza did. Evidence from the *Ricettario Fiorentino* [*Florentine Recipe Book*, 1498] indicates that recipes were also contributed by at least two women: Isabella of Naples (1470–1524), Duchess of Milan, and Isabel Gonzaga (1471–1526),

Fig. 3.3 Frontispiece of the 1565 edition of *I Secreti della Signora Isabella Cortese*

Duchess of Urbino [52]. Adding to this is the fact that during that time, it was not uncommon for women to assist their fathers, brothers, or pharmacist husbands. Not to mention that the exchange of medicinal secrets was a widespread practice. This is evident in *Il Merito delle Donne* [*The Merit of Women*,

1600], a satirical portrait of Venetian society in the second half of the sixteenth century [53]. In this dialogue by Moderata Fonte (pseudonym of Modesta dal Pozzo Zorzi, 1555–1592), seven female friends engage in a debate about the feminine condition. In the second of the two rounds of discussion, they approach the theme of science, offering the reader a small encyclopedia of the scientific knowledge of the time: one by one, they discuss astrology, geography, zoology, botany, mineralogy, and medicine. With great irony, they contrast the effectiveness of some remedies with the futile task of finding a treatment that cures them of the problem of submission to men and, in the case of the latter, that makes them respect women as equals (see excerpt at the beginning of this chapter). They also comment on obtaining gold by transmutation, denouncing the fallacious nature of alchemy and its perverse effects [54].

The book of secrets of Cortese is divided into four parts, the first one consisting of recipes for remedies for diseases such as plague, syphilis, scrofula or gout, and various wounds, as well as preparations for various conditions, from warts to baldness. Curiously, only one of the recipes is exclusively addressed to women: an ointment for stretch marks caused by childbirth [55]. Contrary to what was usual in similar books, for example, in Caterina Sforza's *Experimenti* or della Porta's *Magia Naturalis*, this one does not include secrets and tricks to regain virginity, nor recipes for aphrodisiacs, contraceptives, and abortifacients. The required ingredients are from the three kingdoms of Nature, in more or less complex mixtures. One of the most elaborate recipes is that of a scorpion oil against plague, which, in addition to the arthropod to which the name alludes, requires Malvasia grapes, gentian, saffron, aloe, and rhubarb. The recipe warns of the need for careful handling of the various components and according to a precise ritual, at the risk of the preparation being ineffective or even harmful. The reader is also warned about the influence of the stars on the result. As for the required operations, the following stand out: pulverization, boiling, casting, heating in a water bath, distillation, and drying [56]. In the case of recipes for the plague, the author resorts to a common habit in this type of books: referring to someone notable who has used the preparation and became cured, hoping that the public would trust its effectiveness.

The second part deals with alchemy, being the only one of the four sections that begins with an introduction. The latter consists of a letter addressed to an interlocutor (whom the author refers to as a brother) to whom she intends to reveal the path that leads to alchemical practice. In it, she makes clear her intention to separate herself from the alchemists who "do not tell any truth," and "only tell fables, nonsense, and madness" [57]. The bad reputation that alchemy acquired during the Renaissance, thanks to the proliferation of imposters, often led its adherents to insist on the distinction between good and bad alchemists. The author of *I Secreti* did not spare her criticism, even denouncing renowned authors such as Geber (*c.*721–*c.*815), Ramon Llull (*c.*1232–*c.*1315), and Arnau de Vilanova (*c.*1240–1311), whose works she says to have studied in vain for more than three decades. Anyway, despite her disillusionment with the alchemists, she maintains that, thanks to the providential help of God, she was able to find the right path. This type of reference to divine intervention was also common in alchemical texts from the time of the Counter-Reformation, which, if nothing else, was always a precaution (keeping good relations with the Church could avoid being burned at the stake!). The author also states that the knowledge required to a good alchemist resided in the practical knowledge of metals, especially gold and silver, as well as that of tools and containers necessary for the various operations, some of which are illustrated in the book (Fig. 3.4). The recipes that follow reveal the secrets behind the transformation of base metals into gold and silver or the production of substances such as borax or ammonium chloride [58].

In the third part of *I Secreti*, in addition to recipes for dyeing fabrics and leathers, Cortese teaches how to remove stains, gild books, polish gold, restore fabrics to their original colors, and make glues and paints (e.g., based on blackberry juice) [59]. The fourth and final part is dedicated to the beauty and body care. In a total of 221 recipes, which make up more than half of the book, one finds how to prepare various soaps, waters and perfumed oils, lotions, makeup for the face, creams to depilate the eyebrows, pastes to whiten the teeth, and tinctures to dye the hair and the beard (showing that the book was also aimed at a male audience) [60], in short, a variety of *miraculous* formulas to achieve and maintain a perfect face.

One aspect that is interesting to note is that there are only four parts of the body that deserve attention in the book: the face, the hair, the teeth, and the hands, that is, only those that were visible to others. For the presumable bad body odors, not to mention the infesting parasites (at a time when baths were a rarity), the author gives the recipe for various scented waters, namely, by distilling jasmine or orange blossoms, as well as for preparations such as the *unguento di rogna odorífero*, which was a fragrant ointment for scabies [61].

It is obvious that nowadays these recipes, objectionable in many ways, are nothing more than a mere historical curiosity. First of all, the extravagance of some ingredients, the toxicity of some of them (and the innocuousness of others), as well as the total omission of their quantities are notorious. Examples of substances harmful to health include white lead (basic lead(II) carbonate), used in recipes to lighten the skin [62], and corrosive sublimate (mercury(II)

28   L I B R O   II.

no finiſſimo e ſottile,e parte del corpo reſterà nel fon-
do,per la ſua groſſezza che nõ ual niéte, e tutto quel
che ſerà paſſato per feltro , congela , che ſarà circa.
℥.iij.s.e coſi ſolui e congela tre uolte poi fondi.℥.x. di
argéto fino copellato,e quando ſarà fuſo matti sù.℥.i.
di queſta medicina,e diuenterà tutta medicina, ſimil
mēte fondi borace, cera , e della detta medicina ana.
℥.i.e metti tutto queſto ſopra lib.iij.d'argento uiuo , o
ſopra che corpo tu uorrai,e ſarà argento finiſſimo, ad
ogni giudicio,e coſi ſi farà dell'oro.

E coſi è finito queſto partic olare,ilquale ſi puo fa-
re in quaranta giorni a chi ha buona pratica,e ſa ben
ſollecitare l'opera,ringratiato ſia Iddio.

Queſti ſono i uaſi della detta opera .

Fig. 3.4 Furnace and containers necessary for the practice of alchemy. Illustration from the work of Isabella Cortese (1565 edition, p. 28)

chloride) and metallic mercury, which were components of a facial cosmetic [63]. On the other hand, some procedures are timeless: similarly to what happened in the past, almond milk and lemon juice are still used to whiten hands, eggs and honey are used for face masks, and cloves are for dental products.

The impact of books of secrets on sixteenth-century Italian society—in particular those aimed especially at the female audience—early on attracted the attention of authors of other literary genres [64]. From a wide range of books, *La Raffaella ovvero della Bella Creanza delle Donne* [*Raffaella, or the Beautiful Creation of Women*] stands out. Its author was the humanist, philosopher, astronomer, and science popularizer Alessandro Piccolomini. Published in Venice in 1539, this scathing dialogue stirred up critics and attracted the general public. For some, it was an insult to women; for others, a veritable manual of adultery [65, 66]. (The truth is that Piccolomini was nonetheless an advocate of female education [67].) Addressing very explicitly the commitment of women in the preparation of cosmetics and the use they made of them, *La Raffaella* gives voice to those who considered that artificial embellishment was perverse, being a form of moral corruption. Through the beauty advice of the mature Raffaella to the young and newlywed Margarita (who is urged to find a lover while she is young), Piccolomini ridicularizes the Renaissance manuals of behavior. As a dialogue intended to alert to the moral risks of using cosmetics, *La Raffaella* shares parallels with the *Oeconomicus* of Xenophon (*c*.430–*c*.355 BC), which condemns the use of makeup as nothing more than a trick. Interestingly, it was Piccolomini who translated *Oeconomicus* into Italian shortly after the publication of *La Raffaella*. However, the work of the ancient Greek philosopher had previously served as a model for Leon Battista Alberti's *Quattro Libri della Famiglia* [*Four Books on the Family*], written between 1433 and 1434. This work includes the story of a woman who became disfigured by using an arsenic preparation to lighten her face [68, 69]. The dialogue of Piccolomini goes so far as to include detailed recipes for beauty waters, oils, lotions, and teeth whiteners, thus establishing a close association with the books of secrets. Raffaella's own language seems to be taken from one of them: when recommending that Margarita always use the best cosmetics, she says that she will provide her with "most perfect and rare recipes" [70]. As in the *libri di segreti*, including that of Isabella Cortese, the recipes presented in *La Rafaella* have a strong alchemical basis, both in the materials and in the processes used. For instances, in one case, the distillation of mercury in a water bath is required [71]. In this regard, historian Meredith Ray observed that although the book of Piccolomini conveys a clear message of disapproval of women's use of cosmetics, by disfiguring them both physically and morally, the minutiae of the recipes it provides betrays that objective, making the dialogue, per se, a true repository of secrets [72].

La Rafaella is a work that emerged within the framework of the *querelle des femmes* (see Chap. 2), which at the time had gained a new lease of life. The same can be said about *Lettere di Molte Valorose Donne* [*Letters of Many Valorous*

Women], a satire by Ortensio Lando published, also in Venice, 9 years later. In this collection of letters, attributed to 181 women (some of them of real existence and others purely imagined), the author, a humanist, addresses themes that were mainly of interest to women. These included subjects related to female education, marriage and pregnancy, as well as issues of practical nature, namely, with regard to curing diseases and rejuvenation and beautification. For this purpose, Lando resorted to books of secrets [73–75]. It is interesting to note in *Lettere* the dichotomy between secrecy and the dissemination of knowledge (an issue that would be brought to light again in the seventeenth century, by figures such as Francis Bacon—see next chapter). Despite all the secrecy invoked, the women of Lando are shown to be not able to keep a secret. As for the author, if on the one hand he criticizes female chatter, on the other he also does not abstain himself to question the books of secrets and the inherent contradiction between secrecy and the dissemination of information [76]. Thus, in one of the epistles of *Lettere*, an aristocrat named Argentina reveals to a friend the recipe of an elixir for multiple diseases, making it clear, however, that only the great friendship that united them had led her to share such a secret, so precious that not even to her own son she would reveal it [77]. In another letter, a woman, Livia Beltrama, scolds another one, Aria della Rovere, saying: "all the harm that befell you is due to your inability to keep secret what you were told confidentially." Not yet satisfied, she still throws at her: "You would have burst if you had not given birth to that insignificant little secret" [78]. In this entire process of exchanging recipes and secrets, knowledge works as a bargaining chip for obtaining multiple favors, even social status [79]. The more secret a recipe was, the more profitable it would be, even if it were to be exchanged for another good secret. This is illustrated by Argentina, Countess of Rangona, who tells her interlocutor: "I want to repay the secret, which in the past you sent me, with another secret that is no less capable of protecting the human body from many diseases" [80]. In his collection of epistles, Lando does not miss the opportunity to disdain alchemy, which, thanks to the printed books of secrets, at the time attracted a growing number of adherents, as well as crooks. If, on the one hand, the figure of Giulia Gonzaga, a sixteenth century Italian noblewoman, is highly critical concerning the alchemical desire for transmutation, on the other hand, *Lettere* also parodies the practical and medicinal applications of alchemy [81]. This is the case of a letter attributed to Isabella Sforza (1503–1561), in which the preparation of an elixir based on complicated alchemical ingredients such as "silver foam" and "gold foam" is described. After multiple distillations, it could cure leprosy, remove stains from clothing, improve eyesight, and give eternal youth: "The most miraculous water that both man and woman have ever prepared" [82].

3.3   Closets and Cabinets

The spirit of the Renaissance spread across Europe, carrying with it the idea of the importance of educating aristocratic girls. England, influenced in part by the fact that Thomas More (1478–1535) was the father of three daughters, whom he raised in the same manner as his only son, had a word to say in this respect. For the avoidance of doubt, in his renowned work *Utopia* (1516), a satirical portrayal of an ideal and nonexistent society, the great humanist advocated for equal access to education regardless of gender, age, or social class. Margaret, the most accomplished of his offspring, achieved the remarkable feat of translating the *Pater Noster* of Erasmus of Rotterdam (1466–1536) from Latin to English at the age of 19. This translation, published in 1524, included an introduction in defense of female education by the humanist and educator Richard Hyrde [83]. Erasmus himself, in his colloquium *Abbatis et Eruditae* [*The Abbot and the Erudite*, 1524], wherein he places Magdalia, a cultured lady, in dialogue with Antronio, an ignorant abbot, championed the paramount importance of reading for women, as it would enrich their knowledge [84]. Nevertheless, Erasmus maintained an ambivalent stance regarding female education, believing that instructing women did not necessarily enable independent thinking but rather served as a foundation for their moral development, making them good wives and mothers [85].

Among the sixteenth-century figures who advocated for female literacy in England, Catherine of Aragon (1485–1536), the first of Henry VIII's seven wives, deserves mention. In the *Coronation Suite*, a collection of poems dedicated to the young monarch, Thomas More highly praised the queen consort. A great patron of Renaissance humanism, she appointed her countryman Juan Luis Vives (1493–1540) as the tutor to her daughter, the future Mary I of England. Vive's *De Institutione Feminae Christianae* [*Instruction of a Christian Woman*, 1524], dedicated to Catherine of Aragon, was primarily a book of conduct but became one of the most influential treatises of the sixteenth century on the education of women. Vives believed them to be more intelligent than men [86]. Originally in Latin, this work was translated into English in 1540 and went through a total of 40 editions in Spanish, English, Dutch, French, German, and Italian. Vives, similar to Erasmus, his colleague in Louvain, prohibited women from reading certain books, especially those with love themes. He believed that study not only countered idleness and vanity in women but also preserved their chastity and developed the moral qualities necessary for their roles of wives and mothers [4, 87].

The efforts of advocates for female education in England bore fruit: throughout the sixteenth century, there was a growing trend among upper-class families to prioritize the education of their daughters. The debate on the value and role of women in society, initiated over a century ago in France by Christine de Pisan (as discussed in Chap. 2), was not forgotten. In the apartments of Queen Elizabeth I, the daughter of Henry VIII and Anne Boleyn, and one of the most cultured women of her time, there hung a tapestry inspired by *The Book of the City of Ladies*. It was, however, a copy, as the original had been created in 1513 for Margaret of Austria, Duchess of Savoy, and unfortunately, none have survived [88].

In strong parallelism with the Italian books of secrets aimed at a female audience, the latter half of the sixteenth century and the early seventeenth century saw the proliferation of practical manuals in England. These books, primarily comprising recipes and advice for domestic activities, carried titles such as *The Widdowses Treasure* (1595) or *Delightes for Ladies* (1600). Interestingly, they were all authored by men [89]. The great success of these manuals can be attributed to the significant roles that women played within their families. They were responsible for tasks ranging from food preparation and the preservation of food to maintaining family health and producing various materials, including cleaning agents, cosmetics, perfumes, candles, glues, paints, and varnishes. While these books had deep roots in their Italian counterparts, they were also influenced by English manuals of husbandry and dietetics. In the sixteenth century, Galenic ideas continued to dominate the understanding of health, emphasizing the importance of proper food in disease prevention. One of the notable manuals on this topic, *Dietary of Health* (1542), was authored by the Scottish physician Andrew Borderer. However, by the end of the sixteenth century, the influence of Paracelsus's iatrochemistry, which combined chemistry with medicine, began to emerge in England. This ushered in a new trend of treating diseases with remedies rather than solely focusing on disease prevention through diet. Another significant development was the publication of the *Pharmacopoea Londinenses* [*London Pharmacopoeia*] in 1618, under the auspices of the Royal College of Physicians. This publication aimed to regulate the preparation of drugs. The translation of this work into English in 1649 by the physician and herbalist Nicholas Culpeper, who was also the author of *The English Physician* (1652), played a crucial role in disseminating the scientific principles underlying it [90].

Within this context, the first three practical books authored by women appeared in England in the mid-seventeenth century: *A Choice Manual of Rare and Select Secrets* (1653) by Elizabeth Gray, *The Queen's Closet Opened* by the monarch consort Henrietta Maria, and *Natura Exenterata* [*Eviscerated*

Fig. 3.5 *Natura Exenterata* (1655), by Alethea Talbot, with the portrait of the author

Nature, 1655] by Alethea Talbot (Fig. 3.5). While the first two contain recipes of medicinal and domestic culinary applications, the third, as indicated by its subtitle—"secrets digested into receipts, fitted for the cure of all sorts of infirmities"—is almost exclusively dedicated to the preparation of remedies. These remedies, though often involving common alchemical techniques, were prepared in the kitchen, which is why they became known as *kitchin-physick*, sometimes referred to as medical cookery [91, 92].

Natura Exenterata, which from the (al)chemical point of view is the most interesting of the three, begins with a preface in the form of letter, written by someone who signs as Philiatros. The latter not only defends that medicinal recipes are an important form of knowledge, as, in perfect harmony with the emergence of modern science, raises objections to Aristotelian epistemology. While predominantly comprised of therapeutic recipes, the book also includes brief sections on various topics such as dyeing, lace and knitting, horse breeding, distillation, and perfumery, among others. The remedies, along with the material in subsequent sections, are drawn from various sources, with the authors indicated at the beginning of the book. It also features a small section entitled *Chymical Extractions*, which serves as an illustrative example of Paracelsian iatrochemistry. This section contains recipes for a wide range of quintessences, spanning from heavy metals to plant-derived products, as well as instructions for the preparation of an *aurum potabile* (drinking gold). It also provides tables with alchemical symbols and abbreviations for different measures. In general, this book distinguishes itself by emphasizing the importance of precise measurements and the rigor of instructions, in accordance with the standards of the *London Pharmacopoeia*. These characteristics, along with Philiastro's introductory comments, set *Natura Exenterata* apart from other works of its kind [93].

Elizabeth Gray (1582–1651) and Alethea Talbot (1585–1654) were, in fact, sisters. They both served as ladies-in-waiting to Anne of Denmark, the wife of James I of England, and later to Henrietta Maria of France, the wife of Charles I, with whom they shared a strong friendship. When the Civil War erupted, Elizabeth Gray even accompanied the monarch into exile in the Netherlands. Alethea, who was the goddaughter of Elizabeth I, was married to Thomas Howard, Earl of Arundel, a renowned patron of the arts and an avid collector. (Some of the famous Greco-Roman marbles and sculptures he acquired can now be seen in the Ashmolean Museum, Oxford.) The Countess of Arundel, recognized as a literate woman—John Florio dedicated his English translation of the third volume of Montaigne's *Essays* to her in 1608—was also renowned for her knowledge of natural philosophy, alchemy, and medicine. In 1633, upon acquiring Tart Hall, the villa where she had long resided in London, she embarked on a remodeling project. This included the construction of a properly equipped kitchen for conducting experiments and chemical preparations in line with her interests, as described in her work *Natura Exenterata* [94].

The connection of the two sisters to Queen Henrietta Maria made it possible for them to come into contact with numerous figures of English intellectuality. Among those figures was Sir Kenelm Digby (1603–1665), who, in addition to serving as the queen's chamberlain, was a renowned natural philosopher. He admired the modern physics of Galileo and the rationalism of Descartes, all the while practicing *chymistry* and believing in the occult. Sir Kenelm Digby was also one of the founding members of the Royal Society. Apart from his contributions to natural philosophy, he authored books of secrets and recipes, including a cookbook titled *The Closet of the Eminently Learned Sir Kenelme Digbie Kt. Opened* (1699) [95]. From Queen Henrietta Maria's entourage, there was also her physician, Théodore de Mayerne (1573–1655), a Genevan who had participated in drafting the *London Pharmacopoeia*. Additionally, there was the diarist, landscape artist, and naturalist John Evelyn (1620–1706), who shared interest in anatomy, physiology, and *chymistry* among his many pursuits. He eventually became a member of the Royal Society [96].

According to historian Lynette Hunter, the commitment of these aristocrats to writing books of recipes with a strong medicinal focus may have corresponded, at least partially, to a sense of duty toward the community that emerged within the English nobility after the dissolution of the monasteries, carried out by Henry VIII between 1536 and 1541. One should keep in mind that abbeys and monasteries had once been privileged places for providing medical care [91]. The involvement of these seventeenth-century women in such activities raises questions about their intellectual preparation. In truth, by the end of Henry VIII's reign, with examples like the daughters of Thomas More (mentioned above), the education of young noblewomen had improved. They received a humanist education, with instruction primarily in English since Latin had fallen into disuse. Alethea and Elizabeth Talbot might have received such an education, which was later enriched with knowledge of herbalism, alchemy, and iatrochemistry.

The English books of secrets achieved enormous success. The term "closet," often found in their titles, could refer to various meanings. It could designate a pantry, with an obvious connection to the ingredients and preparations resulting from the recipes. Alternatively, it could refer to a small room for retreat and privacy or even a space for storing curiosities and valuable belongings, including secrets. "Cabinet" is another term frequently seen in the titles of works of this genre, evoking something reserved, personal, and secret [97, 98]. Hannah Woolley (1622–c.1675), who is considered the first English woman to make a livelihood from writing domestic manuals (for this reason, she has already been nicknamed the "Martha Stewart of the seventeenth century which" [99]), even used both words in her *The Queen-like Closet or Rich Cabinet Stored with All Manner of Rare Receipts*, published in 1670.

As the publishing market for these books expanded and literacy rates increased, both men and women showed a growing tendency to compile their own recipes. These handwritten collections served as practical home manuals of knowledge and were often passed down to the next generation, who would then add to them [100]. An excellent example of such compilations is *The Book of Recipes of Sarah Wigges* (c.1616), housed in the Wellcome Library of the Royal College of Physicians in London. This manuscript, authored by an English housewife from the 1600s, spans over four hundred pages and includes recipes for cooking, confectionary, medicines, cleaning products, perfumes, cosmetics, rat poisons, and paints. What sets this manuscript apart from most of its counterparts is the inclusion of recipes that seem more suited for an alchemical practice book. It contains a formula for obtaining the Philosopher's Stone, as well as instructions for preparing the "star regulus of antimony," a step that many alchemists believed was crucial in achieving that goal. The last page of the manuscript presents two recipes: one for puff pastry and the other for producing diamonds—two tasks that, as is well known, are anything but easy [101]!

Researcher Jayne E. Archer, who studied Wigges' manuscript, investigated the public perception of women's involvement in *chymistry* in the seventeenth century. To do this, she drew upon the viewpoints of several prominent figures of the time, ranging from Richard Allestree (1621–1681) to Thomas Vaughn (1621–1666), including Margaret Cavendish, Duchess of New Castle (1623–1673). Richard Allestree, a religious intellectual and dean of the renowned Eton College, upheld the traditional notion that women, being seen as vain and not particularly adept in *chymistry*, were better suited for spending gold than producing it. In contrast, Thomas Vaughn, a *chymist* and philosopher, took the opposite stance, arguing that women possessed great aptitude for its practice, especially in the pursuit of the Philosopher's Stone. Meanwhile, the Duchess of Newcastle, known for her socialite status and natural philosophic interests (as will be discussed in detail in the next chapter), took a middle-ground position. She believed women could be as competent as men in the arts of fire but simultaneously more skillful at spending gold than producing it [102].

References and Notes

1. Moderata Fonte, *Il merito delle donne*, E-text, Liberliber, p. 61; https://www.liberliber.it/mediateca/libri/m/moderata_fonte/il_merito_delle_donne/pdf/il_mer_p.pdf
2. Baldassare Castiglione, *Il libro del Cortegiano*, Einaudi, Torino, 1965, p. 1
3. Phyllis Stock, *Better Than Rubies—A History of Women's Education*, Putnam, New York, 1978, pp. 32–36
4. J. Gibson, "The Logic of Chastity: Women, Sex, and the History of Philosophy in the Early Modern Period", *Hypatia*, 21 (2006) 1–19
5. The condottieri led armies of mercenaries in the service of monarchs and popes.
6. J. de Vries, "Caterina Sforza's Portrait Medals: Power, Gender, and Representation in the Italian Renaissance Court", *Woman's Art Journal*, 24 (2003) 23–28
7. Guy Rachet, *Catherine Sforza—La Dame de Forli*, Denoël, Paris, 1987, pp. 95–97
8. *Ibid.,* p. 127
9. J. de Vries, "Caterina Sforza: The shifting representation of a woman ruler in early Modern Italy", *Lo Sguardo—Rivista di Filosofia*, 13 (2013), 165–181

10. Meredith K. Ray, "Experiments with Alchemy: Caterina Sforza in Early Modern Scientific Culture" *in* Kathleen P. Long (ed.), *Gender and Scientific Discourse in Early Modern Culture*, Routledge, New York, 2010, p. 150
11. Rachet, *Op. cit.* (7), p. 245
12. Abbreviated form of *Gli Experimenti de la Ex.ma S.ra Caterina da Furlj, Matre de lo Inllux.mo Signor Giovanni de Medici*
13. A. Rankin, "Becoming an expert practitioner—Court experimentalism and the medical skills of Anna of Saxony (1532–1585)", *Isis*, 98 (2007) 23–53
14. David Wootton, *A Invenção da Ciência*, Temas e Debates—Círculo de Leitores, Lisboa, 2017, p. 396
15. C. B. Schmitt, "Experience and Experiment: A comparison of Zabarella's view with Galileo's in De Motu", *Studies in the Renaissance*, 16 (1969) 80–138
16. W. Eamon, F. Paheau, "The Accademia Segreta of Girolamo Ruscelli: A Sixteenth-Century Italian Scientific Society", *Isis*, 75 (1984) 327–342
17. Joyce de Vries, *Caterina Sforza and the Art of Appearances: Gender, Art and Culture in Early Modern Italy*, Farnham, Ashgate, 2010, pp. 210–211
18. Meredith K. Ray, "Impotence and Corruption: Sexual Function and Dysfunction in Early Modern Italian Books of Secrets", *in* Sara F. Matthews-Grieco (ed.), *Cuckoldry, Impotence and Adultery in Europe (15th-17th century)*, Routledge, New York, 2014
19. R. Palmer, "Medical botany in northern Italy in the Renaissance", *J. R. Soc. Med.*, 78 (1985) 149–157
20. Meredith K. Ray, *Daughters of Alchemy—Women and Scientific Culture in Early Modern Italy*, Harvard University Press, Cambridge, 2015, pp. 22 and 28
21. Pier Desiderio Pasolini, *Caterina Sforza*, Vol. III, Ermanno Loescher, Roma, 1893, pp. 622–624
22. *Ibid.*, pp. 689–690
23. Castiglione, *Op. cit.* (2), p. 217
24. *Ibid.*, p. 219
25. Pasolini, *Op. cit.* (21), pp. 617–619
26. *Ibid.*, p. 621
27. *Ibid.*, p. 627
28. *Ibid.*, pp. 630–631
29. *Ibid.*, pp. 620 e 782
30. de Vries, *Op. Cit.* (17), p. 211
31. William Eamon, Science and the Secrets of Nature: Books of Secrets in Medieval and Early Modern Culture, Princeton University Press, Princeton, 1994, p. 270
32. Ray, *Op.cit.* (20), p. 44
33. Caroline Desgranges, Jerome Delhommelle, *A Mole of Chemistry—An Historical and Conceptual Approach to Fundamental Ideas in Chemistry*, CRC Press, New York, 2020, p. 6
34. Pier Desiderio Pasolini, *Op. Cit.*, p. 601
35. Ray, *Op. cit.* (10), p. 156

36. Sheila Barker, Sharon Strocchia, "Household Medicine for a Renaissance Court Caterina Sforza's Ricettario Reconsidered", *in* Sheila Barker, Sharon Strocchia (eds.) *Gender, Health, and Healing, 1250-1550*, Amsterdam University Press, Amsterdam, 2020, pp. 139–166

37. According to Ray, *Op. cit.* (20), p. 171, the *Ricettario Galante*, published in 1883, was transcribed from a manuscript existing in the library of the University of Bologna.

38. Eamon, *Op. cit.* (31), p. 4

39. William E. Burns, The Scientific Revolution*: An Encyclopedia*, ABC—Clio, Santa Barbara, 2001, p. 35

40. Giambattista della Porta, *Natural Magick*, Basick Books, New York, 1958, p. VI

41. John Henry, *The Scientific Revolution and the Origins of Modern Science*, Palgrave Macmillan, 2008, pp. 56–57

42. Eamon, *Op. cit.* (31), p. 225

43. Henry, *Op. cit.* (41), p. 59

44. della Porta, *Op. cit.* (40), p. 252

45. Eamon, *Op. cit.* (31), pp. 202–203

46. The subtitle is *Ne' quali si contengono cose minerali, medicinali, arteficiose e alchimiche, e molte de l'arte profumatoria, appartenenti a ogni gran Signora* [*Which include mineral, medicinal, artificial, and alchemical things, as well as much of the art of perfumery, belonging to any great lady*]

47. The consulted edition was the 1565 edition.

48. Bruce Moran, *Distilling Knowledge: Alchemy, Chemistry, and the Scientific Revolution,* Harvard University Press, Cambridge, MA, 2009, p. 61

49. Ray, *Op. cit.* (20), p. 55

50. *Claire Lesage, "La litterature des "secrets" et I secreti d'Isabella Cortese", Chroniques italiennes, Université Paris III,* 36 (1993) 145–178

51. Giovanni Marinello, *Gli ornamenti delle donne*, Francesco de Franceschi, Venetia, 1562, p. xi

52. Raffaele Ciasca, *L'arte dei medici e speziale nella storia e nel commercio Fiorentino, dal secolo XII al secolo XV*, p. 345 da edição de Leo S. Olschki, Ristampa, Firenze, 1927

53. B. Collina, "Moderata Fonte e "Il Merito delle Donne"", *Annali d'Italianistica*, 7 (1989) 142–164

54. Fonte, *Op. cit.* (1), p. 68

55. Isabella Cortese, *I Secreti della Signora Isabella Cortese*, Giouanni Bariletto, Venetia, 1565, p. 11

56. *Ibid*, pp. 6–7

57. *Ibid*, p. 19

58. *Ibid*, p. 42

59. *Ibid*, p. 67

60. *Ibid*, p. 174

61. *Ibid*, p. 167

62. *Ibid*, p. 164

63. *Ibid*, p. 100

64. Ray, *Op. cit.* (20), p. 62

65. Ian Frederick Moulton, "Vagina dialogues: Picolomini's *Raffaella* and Aretino's *Ragionamenti*", *in* Jacqueline Murray, Nicholas Terpstra (eds.), *Sex, Gender and Sexuality in Renaissance Italy*, Routledge, Abingdon, 2019

66. R. Suter, "The Scientific Work of Allesandro Piccolomini", *Isis*, 60 (1969) 210–222

67. S. Plastina, "Politica amorosa e'governo delle donne' nella Raffaella di Alessandro Piccolomini", *Bruniana & Campanelliana*, 12 (2006) 81–94

68. Stock, *Op. cit.* (3), pp. 30–31

69. Ray, *Op. cit.* (20), pp. 62–63

70. Alessandro Piccolomini*, La Raffaella, ovvero della Bella Creanza delle Donne*, Daelli, Milano, 1862, vol. 1, p. 25

71. *Ibid.*, p. 28

72. Ray, *Op. cit.* (20), p. 65

73. M. K. Ray, "Textual Collaboration and Spiritual Partnership in Sixteenth-Century Italy: The Case of Ortensio Lando and Lucrezia Gonzaga", *Renaissance Quarterly*, 62 (2009) 694–747

74. Ray, *Op. cit.* (20), p. 66

75. M. K. Ray, *Writing Gender in Women's Letter Collections of the Italian Renaissance*, University of Toronto Press, Toronto, 2009, p. 73–74

76. Ray, *Op. cit.* (20), p. 67

77. Ortensio Lando, *Lettere di Molte Valorose Donne, nelle quali chiaramente appare non esser ne di eloquentia ne di dottrina alli huomini inferiori*, Giolito, Veneza, 1548, p. 113v

78. *Ibid.*, p. 58r

79. Ray, *Op. cit.* (20), p. 69

80. Lando, *Op. cit.* (77), p. 115

81. Ray, *Op. cit.* (20), p. 70

82. Ray, *Op. cit.* (20), p. 116

83. Stock, *Op. cit.* (3), p. 50

84. Maria de Lurdes Correia Fernandes, *Espelhos, Cartas e Guias—Casamento e Espiritualidade na Península Ibérica 1450-1700*, Faculdade de Letras da Universidade do Porto, 1995, p. 123

85. Leigh Ann Whaley, *Women's History as Scientists—A Guide to the Debates*, ABC CLIO, Santa Barbara, 2003, pp. 52–53

86. Juan Luis Vives, *The education of a christian woman: a sixteenth-century manual—Edited and translated by Charles Fantazzi*, The University of Chicago Press, Chicago, 2000, p. 2

87. Stock, *Op. cit.* (3), p. 51

88. Susan G. Bell, *The lost tapestries of the city of ladies: Christine de Pizan's Renaissance*, University of California Press, Berkeley, 2004

89. The most successful authors of this period were Thomas Tusser (author of *Five Hundred Points of Good Husbandry*, 1557), Thomas Dawson (*The Good Huswifes Jewell*, 1585), John Partridge (*The Widdowes Treasure*, 1595), Hugh Plat (*Delightes for Ladies*, 1600), and Gervase Markham (*The English Huswife*, 1615).

90. M. DiMeo, "Communicating Medical Recipes: Robert Boyle's Genre and Rhetorical Strategies for Print" Howard Marchitello, *in* Evelyn Tribble (ed.), *The Palgrave Handbook of Early Modern Literature and Science*, Palgrave Macmillan, London, 2017, p. 212

91. L. Hunter, "Women and domestic medicine: Lady Experimenters, 1570-1620", *in Women, Science and Medicine: 1500-1700—Mothers and sisters of the Royal Society*, (Lynette Hunter, Sarah Hutton, eds.), Sutton Publishing, Stroud, 1997

92. L. Lunger Knoppers, "Opening the Queen's Closet: Henrietta Maria, Elizabeth Cromwell, and the Politics of Cookery", *Renaissance Quarterly*, 60 (2007) 464–499

93. Elizabeth Spiller, *Seventeenth-Century English Recipe Books: Cooking, Physic and Chirurgery in the Works of Elizabeth Talbot Grey and Alethea Talbot Howard*, Routledge, Abingdon, 2016, pp. xxxv–xxxvii

94. J. Rabe, "Mediating between Art and Nature: The Countess of Arundel at Tart Hall", *in Sites of Mediation Connected Histories of Places, Processes,and Objects in Europe and Beyond, 1450–1650*, (Susanna Burghartz, Lucas Burkart, Christine Göttler, eds.), Brill, Leinden, 2016, cap. 7

95. J. F. Fulton, "Sir Kenelm Digby, F.R.S. (1603-1665)", *Notes and Records of the Royal Society of London*, 15 (1960) 199–210

96. E. S. de Beer, "John Evelyn, F.R.S. (1620-1706)", *Notes and Records of the Royal Society of London*, 15 (1960) 231–238

97. L. Hunter, "Cookery books: a cabinet of rare devices", *Petits Propos Culinaire*, 5 (1980) 19–20

98. V. Domínguez-Rodríguez, A. González-Hernández, "Remedies for headaches in *A Closet for Ladies and Gentlewomen* (1608)", *Headache*, 51 (2011) 632–636

99. "Early Modern Martha" A mini-podcast series on the life and work of Hannah Woolley;   https://anevolvingworkinprogress.com/2021/12/07/early-modern-martha/

100. E. Leong, "Collecting Knowledge for the Family: Recipes, Gender and Practical Knowledge in the Early Modern English Household", *Centaurus* 55 (2013) 81–103

101. J. E. Archer, "Women and chemistry in early modern England: the manuscript receipt book (*c.* 1616) of Sarah Wigges", *in* Kathleen Perry Long (ed.), *Gender and Scientific Discourse in Early Modern Culture*, Routledge, London, 2010, pp. 191–192

102. *Ibid.* pp. 193–195

4

Atomists and *Femmes Savantes*

Small Atoms of themselves a World may make As being subtle, and of every shape: And as they dance about, fit places find, Such Forms as best agree, make every kind.

MARGARET CAVENDISH, *A World Made of Atoms* (1653) [1]

As a reaction to the Renaissance, throughout the seventeenth century, mathematics replaced magic in interpreting natural phenomena, and mechanical philosophy gradually supplanted organicism. These changes paved the way for the dissemination of atomism, which had originated in Greece in the fifth century BC but had fallen into oblivion until it experienced a resurgence starting in the fifteenth century.

Similar to what had occurred in universities, scientific academies—founded in the second half of the 1600s and inspired by monastic models—also barred women from entry. Consequently, women's engagement with natural philosophy remained confined to informal intellectual circles.

The central figures in this chapter include the pious Lady Ranelagh, who was Robert Boyle's inseparable sister; the extravagant Duchess of Newcastle, a natural philosopher active in diverse literary genres; Christina, Queen of Sweden, who grappled with existential and metaphysical questions; Marie Meurdrac, the first female author of a chemistry book; Madame Fouquet, who also left written work; and Joanna Stephens, known for developing an acclaimed method for disintegrating bladder stones. Reflecting the iatrochemical trend of the time, which applied chemistry to medicine, the

© The Author(s), under exclusive license to Springer Nature Switzerland AG 2024
J. P. André, *Sisters of Prometheus*, https://doi.org/10.1007/978-3-031-57136-7_4

contributions of almost all these seventeenth century women were nonetheless part of the tradition of *kitchin-physick*.

4.1   The Triumph of the Four *M*

Matter, motion, mathematics, and mind were four key concepts of the seventeenth-century philosophy. Building upon the first two concepts—matter and motion—the French philosopher, mathematician, and physicist René Descartes (1596–1650), who regarded reasoning as the basis of knowledge, presented one of the first fully mechanistic views of the universe. This perspective contrasted with the magical and organicist conceptions that had dominated since the Renaissance (see Chap. 2).

Considered one of the most influential books in the history of modern philosophy, his *Discourse on the Method* (1637) was decisive for the commencement of science as we understand it today. Skepticism was not new, but Descartes, in his search for the fundamental truths of reality, distinguished himself from his predecessors and contemporaries by defending a (deductive) methodology whose first step was to question everything that could offer doubt—the so-called method of "Cartesian doubt." Only in this way would it be possible to evaluate the world from a new perspective, free from any preconceived notions—exactly the opposite of Aristotelian scholastic philosophy. Similar to Galileo Galilei, 32 years his senior, Descartes believed that mathematics was the language in which the Book of Nature was written [2]. He was the creator of the so-called Cartesian coordinate system [3].

Despite conceiving that the universe worked like a machine, Descartes did not fail to recognize that its creation was due to a superior being—just as the construction of a watch requires a watchmaker. God had first created matter, and then, before leaving the universe working on its own, He had endowed it with motion. In his works *Principles of Philosophy* (1644) and *The World* (written between 1629 and 1633, but only published posthumously, in 1664), Descartes declared that vacuum and atoms did not exist: the universe was a *continuum* of matter, a *plenum* composed of a network of vortices, formed by particles of subtle matter. It was these whirlpools that dragged the planets around the sun. Not conceiving that there were actions at a distance, as Newton would later demonstrate with the gravitational theory, he understood that matter only moved when pushed (like billiard balls) or pulled, obeying the laws of mechanics. As for the bodies of men and animals, Descartes considered them to be machines, with animals being true automata as they lacked consciousness. His famous assertion *cogito, ergo sum* ("I think, therefore I am"), expressed in the *Discourse on the Method*, contained the idea that he could doubt the existence of the material world (of which the body is a part), but not the existence of himself as a thinking reality; hence, his thoughts belonged to a non-spatial domain, distinct from matter. This is the so-called mind-body dualism, which would stir up a hornet's nest of philosophical debates.

Descartes kept God at the top of creation, but several of his followers, for foreseeing possible atheistic distortions to his mechanistic doctrine, endowed it with contours that unequivocally asserted the existence of the Creator. One of the most renowned was the Irishman Robert Boyle (1627–1691), who, despite being associated with the gas law named after him (at a given temperature, the pressure of a gas is inversely proportional to its volume) and believing that the world was made up of small corpuscles, has argued that the practice of natural philosophy was intended to reveal the splendor and wisdom of God [4–7].

Despite the limited empirical (sensory) foundation of atomism (Boyle himself acknowledged, in the 1660s, that he had to concede that "the intelligible [corpuscular] philosophy... seemed to have not employed, let alone produced, any set of experiences up to that point"), the new natural philosophy would be seduced by the idea of describing the phenomena of Nature through particles and motion [8]. This was the same Boyle who asserted that the purpose of experimentation was to uncover the hidden or occult mechanical causes of all effects [9].

As secular supporters of religious orthodoxy, the new natural philosophers aimed to maintain their dominion closed to the female sex. Just as the interpretation of the Scriptures had traditionally been the work of men, they believed that the mathematical decoding of Nature (considered another divine book) should be an exclusively male task. Consequently, women were a priori excluded from the emerging academies. According to some historians, such discrimination was in complete harmony with the monastic model advocated by many followers of the new philosophy in these recent associations [10]. An example of this was the pioneer Accademia dei Lincei (to which Galileo Galilei and Giovanbattista della Porta belonged—see Chap. 3), founded in Rome in 1603 by Prince Federico Cesi, whose family had strong ties to the Church. Cesi advocated that natural philosophy should be developed in brotherhoods of men who worked and lived in the same place and who were single and chaste. In the original project of Cesi, those men were obliged to avoid the "attractions of Venus," the "venereal lust," the "tempting lust," the "low passions of the body," the "carnal impulses," the "libidinous arousals," and the "insane desires of the body." Only a pure mind would have within its reach the discovery of true knowledge, so all contacts with women were harmful. No maid should even be hired (unless she was old and ugly!); the cleaning of the place and the care of the members of the brotherhood would be entrusted to a male servant (they were also compelled to keep away from "scandals with boys"!) [11].

John Evelyn (1620–1706), one of the founders of the Royal Society of London, advocated for an equally monastic model for the latter, in which his confreres would wear habits, pray, take communion, fast, and live in cells. In reality this did not occur, but in a similar way to what happened with French mechanists—among whom, besides Descartes, were clerics such as Marin Mersenne (1588–1648) and Pierre Gassendi (1592–1655)—many of its members made a point of demarcating themselves from everything that could be considered heretical, which included the female sex. Some of them, such as Robert Boyle, took a vow of chastity. His assistant, Robert Hooke (1635–1703), himself one of the great physicists of his time, and later secretary of the Royal Society, took a vow of celibacy (but had sexual relations with the maids and, from 1676, with his niece, the daughter of a brother!) [12]. In turn, Isaac Newton (1643–1727), who presided over the Society for more than two decades, died a virgin. Henry Oldenburg (c.1618–1677), the first secretary, was peremptory in stating that he wished to "promote a Masculine Philosophy [...] through which the Mind of Man can be ennobled with the knowledge of Solid Truths" [13]. His assertion would prove to be of unshakable solidity for nearly three hundred years, as the Society's only concessions to the female sex were, as will be seen, the visit of the Duchess of Newcastle and a woman's skeleton in the anatomy collection [14].

While it is true that Descartes, when defending that mind and body functioned separately, had provided arguments that could contest the Aristotelian conviction that women were intellectually inferior to men, it is also no less true that his philosophy presented no defense or praise of the feminine sex. However, the French philosopher corresponded with two intellectual women: Elisabeth of Bohemia (to whom he dedicated his *Principles of Philosophy*, published in 1644) and Christina of Sweden, as it will be seen in this chapter.

4.2   Such a Sister, Such a Brother

It is possible to trace the practice of *chymistry* (see Chap. 2) by English aristocrats since the Elizabethan period (Elizabeth I might have been a practitioner herself), namely, by figures such as Margaret Clifford (1560–1616), Countess of Cumberland, and Mary Sidney Herbert (1561–1621), Countess of Pembroke [15]. Concerning the former, who compiled a set of (al)chemical recipes [16, 17], her own daughter, Lady Anne Clifford, said:

She was a lover of the study and practice of alchimy, by which she found out excellent medicines, that did much good to many; she delighted in the distilling of waters, and other chymical extractions, for she had some knowledge in most kinds of minerals, herbs, flowers and plants [18]

A few decades later, the Anglo-Irish Dorothy Dury (161–1664) dedicated herself to extracting essential oils of vegetable origin. In 1649, as she was planning to open a home remedies shop (possibly for financial reasons), she sought to learn the art of distillation [19]. In 1654, she worked with at least one of the Boate brothers, Arnold and Gerard, on the "Paris chemistry" (both brothers were physicians and part of Samuel Hartlib's circle—see below) [20, 21]. It was in France, at the beginning of the seventeenth century, that the so-called *rational chemistry* emerged, strongly associated with the idea of utility. Such chemistry was above all iatrochemistry, that is, chemistry intended for medical use. The first French treatise on *rational chemistry* was the influential *Tyrocinium Chymicum* [*Chemical Training*, 1610] by Jean Béguin (1550–1620), which had a total of over 40 editions in Latin, French, and English. Béguin, a pharmacist like most French chemists at the time, had established a school in Paris, and it was with his students in mind that he wrote the book [22].

To the names of these English noblewomen, those of the authors of books on *kitchin-physick* should be also added, particularly Alethea Talbot, as mentioned in the previous chapter. However, it was Katherine Jones, Lady Ranelagh (1615–1691), who actually was a niece (by affinity) of Dorothy Dury, who reached the zenith of the practice of *chymistry*. The daughter of Richard Boyle, the very wealthy Earl of Cork, the "incomparable" Lady Ranelagh was one of the exponents of female intellectuality in the British Isles in the seventeenth century. It is not known what education she received, but judging by the most famous of her 14 brothers, Robert Boyle, and, ultimately, by her father's social and economic position, one would imagine it had been thorough. At just 15 years old, she married Arthur Jones, future Viscount of Ranelagh, but his debauchery did not help the couple's happiness, despite the four children they had. After years of constant back and forth between Ireland and London, in 1642, finally separated from her husband, she took up residence in the English capital. A devout Protestant, kind and prone to compromise (reflecting the diversity of the family's religious and political options: monarchists, parliamentarians, Catholics, Protestants and Puritans), Lady Ranelagh was admired and respected by all rulers from Oliver Cromwell to the royal couple William and Mary. The intellectual circles in which she moved, notably the Great Tew Circle, the Invisible College, and the Hartlib Circle, included physicians and natural philosophers such as Henry

Oldenburg, John Evelyn, Robert Hooke, Kenelm Digby, or John Beale (eventually they all would be members of the Royal Society of London). In the field of letters, she crossed her paths with figures such as Edmund Waller, Andrew Marvell, and John Milton, whom she chose to tutor her son Richard [23].

The Hartlib Circle was an intellectual correspondence network formed in London around 1641. Centered on Samuel Hartlib, John Dury (Dorothy Dury's second husband), and Jan Amos Comenius, its mission was to contribute to the socioreligious reform of the nation. With the motto "for the public utility, for the glory of God," it remained active until 1661, mirroring the religious, political, and scientific environment experienced during the Civil War (1642–1649) and the Interregnum (1649–1660). Its plans for social and educational development were supported by the Protestant (mostly Puritan) impulse to transform England into a society in which prevailed the value of compassion. These three men were not only interested in the new discoveries of natural philosophy, but they also shared the conviction that all useful knowledge should be put at the service of society, according to the philosophical ideas of Francis Bacon (see below) [24]. In an unusual fashion for the time, the Hartlib circle included some women, among which Lady Ranelagh and Dorothy Dury stood out [25, 26]. In addition to their common passion for chemistry, aunt and niece came to outline an education plan for young women, which, however, would never be put into practice [27, 28].

In the late 1640s, Lady Ranelagh's house in Queen Street was said to have been the meeting place of the Invisible College, to which belonged, among others, her brother Robert Boyle and the famous architect and astronomer Christopher Wren. This group, thus coined by Boyle, would eventually be the embryo of the Royal Society for Improving Natural Knowledge, founded in London in 1660 and known today only as the Royal Society. These men, for whom Aristotelianism had failed, believed that it was possible to change the world for the better using the knowledge obtained through experimentation and the rational exploration of natural phenomena. Their spiritual mentor was Francis Bacon (1561–1626), an English philosopher and statesmen who, in his *Novum Organum* [*New Organon*, 1620], had presented a new methodology for natural philosophy, which today is called the scientific method. The title of his work alludes to *Organon*, the treatise of Aristotle on logic and syllogism (see Chap. 1). Unlike the ancient philosopher, Bacon, who believed that the laws of Nature could only be unveiled by collecting and organizing large amounts of data, advocated an inductive approach in which explanations were inferred from unbiased observations [29].

Katherine and Robert Boyle were two inseparable siblings, which may be explained by the circumstance that she was already 15 and he was only three when they were orphaned. To tell the story of Lady Ranelagh, especially from 1644 onward, is in large part to tell the story of Robert Boyle. That year, at just 17 years old, having returned from a 5-year *Grand Tour* of mainland Europe, he paid a visit to his sister in London. From then on, he lived in Stalbridge House, a mansion owned by the family in the countryside, in Dorset. Thus began an uninterrupted and deep fraternal bond, characterized by the sharing of intellectual aspirations and collaboration in projects that included religion, chemistry, and medicine. Lady Ranelagh's constant support of her brother's philosophical and theological ambitions undoubtedly contributed to his notability. She was instrumental in introducing him to Samuel Hartlib and other important figures in the latter's circle, namely, Henry Oldenburg.

During the first 5 years spent at Stalbridge, Boyle devoted himself mainly to the themes of morals and ethics. It was not until 1649 that he began to seriously explore chemistry (still with a great alchemical bent), which at that time also began to arouse a special interest in Lady Ranelagh, as the correspondence between the two brothers reveals. As historian Michelle Dimeo notes, it is difficult to establish which of them was first seduced by chemistry [30]. Documents in the Hartlib archive attest that between 1643 and 1648 Lady Ranelagh was mostly involved in discussions of political, religious, and moral nature and that it was at the end of the decade that her interest in that branch of knowledge intensified. The two siblings discussed experiences and exchanged recipes. It was Lady Ranelagh herself who, in 1647, helped her brother to equip his first chemical laboratory at Stalbridge. The furnace, unfortunately, arrived badly damaged, as the very disgruntled Boyle related to her in a letter dated March 6th:

> That great earthen furnace, whose conveying hither has taken up so much of my care, and concerning which I made bold very lately to trouble you [...] has been brought to my hands crumbled into as many pieces, as we into sects; [...] I see I am not designed to the finding out the philosophers stone, I have been so unlucky in my first attempts in chemistry. My limbecks, recipients, and other glasses have escaped indeed the misfortune of their incendiary, but are now, through the miscarriage of that grand implement of Vulcan, as useless to me, as good parts to salvation without the fire of zeal [31]

Lady Ranelagh not only encouraged her brother to pursue his chemical investigations but also stimulated him to write essays, such as *An Epistolical Discourse*, drawn up between 1647 and 1649. In this manifesto against secrecy in natural philosophy, Boyle discusses the moral obligations of natural philosophers, in complete alignment with Lady Ranelagh's commitment to promoting Baconian ideology. In the mid-1650s, after discussing with his sister the possibility of moving to Oxford, she went to that city herself to arrange for his lodging at the home of the pharmacist John Crosse on High Street. Despite the lack of comfort found, Lady Ranelagh recognized the great advantage of having a laboratory there, where her brother could work [32, 33].

Many of Boyle's chemical interests coincided with those of his sister, which, according to correspondence from the Hartlib circle, in the late 1650s came to include the search for the Philosopher's Stone [30]. However, the greatest passion of Lady Ranelagh was medicine, a field in which she acquired a great reputation after the Restoration. She was often called upon to treat people of high social status, especially when conventional doctors could no longer do anything. In the mid-seventeenth century, the general trend was to force women out of clinical practice so that it was left solely to formally educated (male) practitioners. In view of this, women were left to work as healers, practicing a domestic medicine that was protected by the legislation that applied to apothecaries. Within the aristocracy, this practice (unpaid) was highly appreciated and was even seen as a form of charity [34].

Lady Ranelagh's recipes and correspondence reveal a variety of influences: to the old Galenic doctrines of humoral imbalance and diet, she added the new chemical remedies, following the Paracelsian tendency [35]. Her brother himself paid tribute to her in the second part of his *Usefulness of Experimental Natural Philosophy* (1663), dedicated to his nephew Richard Jones, her son. In this work, he mentions some of his sister's remedies, among them the "*Laudanus opiatum* of Helmont," "Sir Walter Rawleighs Cordial" and the "Colcotharine flowers" [36]. In this treatise, Boyle implies that they have worked together in some chemical-medicinal projects, namely, in the therapeutic use of *Ens veneris* (the "Colcotharine flowers") [37], referred to as a "specific potent for rickets", with which Lady Ranelagh allegedly cured hundreds of children [38]. It was, in fact, a copper compound that the Flemish iatrochemist Jan Baptist van Helmont (1580–1644) had cryptographically described and that Boyle prepared and studied with the alchemist George Starkey [39]. It is worth mentioning that van Helmont was one of the important figures in the transition from alchemy to modern chemistry. Although a follower of Paracelsus, he has been an experimentalist. He abolished the Aristotelian four elements, as well as the theory of the four humors, and defended the existence of only two elements: water and air. He coined the term "gas," from the Greek *chaos*, meaning chaos.

In 1660, with the creation of the Royal Society, the Circle of Hartlib—Lady Ranelagh's main intellectual network for about two decades—was dissolved. This was partly due to the fact that some of its members, namely, Boyle, were among the founders of that institution and others were elected its members. As a woman, Lady Ranelagh could not have any linking to the Royal Society, but she kept in touch with several of her former confreres, namely, with John Beale and Henry Oldenburg. Additionally, she created new relationships, namely, with John Evelyn and Robert Hooke. Despite all these changes, she continued encouraging her brother and reading his manuscripts critically, believing strongly in the benefit his work brought to society. By this time, Boyle was increasingly involved in writing on the subject of natural philosophy. His masterpiece, *The Sceptical Chymist*, a long dialogue on the nature of the elements and their number, dates from 1661. In this work, the Irish chemist considers the elements as primitive and simple bodies because they are not made of any others or of each other. They are the ingredients that make up all the so-called "perfectly mixed bodies" (as opposed to mechanical mixtures) and into which the latter can be resolved. Although some authors find here the first modern definition of a chemical element, this is not consensual, because Boyle did not cease to consider that the different elements were constituted by a primordial matter and that their properties were due to the shapes and movements of the particles of that matter [40].

It is noteworthy that the founding year of the Royal Society coincided with the restoration of the English monarchy, under Charles II (who, 2 years later, would marry Catherine of Braganza). This marked the end of a turbulent period, including the execution of Charles I in 1649, Oliver Cromwell's dissolution of Parliament in 1653, and his self-proclamation as protector of the Republic of England, Ireland, and Scotland.

In 1668, Boyle moved definitively to London, to Pall Mall, door to door with his sister (apparently on the site where the Royal Automobile Club now stands). Once again, she made sure that her brother had adequate space for his investigations, eventually hiring Robert Hooke as the architect for the house improvement and laboratory expansion. For the last 23 years of the celibate Boyle's life, the two siblings regularly dined together, shared information on medical diagnoses and remedies, and provided each other with chemical reagents. Their passion for natural philosophy was matched only by their strong religiosity, which, in fact, was the great driving force of their lives. Union in life and death seemed to be the motto of this fraternal duo: they died in 1691, within a week of each other. In the eulogy of Robert Boyle, who died after his sister, the Bishop of Salisbury also referred to Lady Ranelagh, noting that she had been a unique female figure, who, thanks to her knowledge, her charity, her humility, and religiosity, had been prominent in the developments of the nation for five decades. He concluded his speech with the words: "Such a sister became such a brother" [30].

The work that has come down to us from Lady Ranelagh is limited to manuscripts. Respecting the feminine conventions of modesty and reserve, she deliberately avoided publishing. In his dedication to his "Dearest Sister" in *Occasional Reflections Upon Several Subjects* (1665), Robert Boyle mentions this characteristic of hers. After apologizing for offering "trifles to someone who deserves the noblest productions of [...] wit and eloquence," he suggests that had her modesty not confined her pen to excellent letters, male minds would have envied such a writer [41]. However, this comment about the circumscription of his sister's writing to the epistolary genre is not entirely accurate. While it is true that she presented most of her intellectual output in the form of letters, Lady Ranelagh was also active in other forms of writing. Documents from the Hartlib archive reveal that she wrote and disseminated recipes, specifically for medicinal applications [42]. Additionally, she authored two theological treatises, intended for limited circulation (only one of which survives). There are also three manuscripts of recipe books that have traditionally been attributed to her. The one in the Wellcome Library in London has long been attributed to the Boyle family. It is a book of recipes for domestic preparations, encompassing both therapeutic and culinary uses (see Chap. 3). Among its contents is the recipe for the "Spirit of roses in the manner of my brother Robert Boyls [*sic*]" [43]. In a recent analysis, DiMeo revealed that the work has been written in three different handwriting styles, none of which belong to Lady Ranelagh herself. This researcher suggested that the first author of the manuscript has been her sister-in-law, Margaret Boyle, Countess of Orrery, who used to refer to her brother-in-law, Boyle, as her brother [44]. Additionally, in the British Library, there exists a manuscript book titled *My Lady Ranelagh's Choice Receipts*. This compilation primarily contains chemical recipes and is believed by DiMeo to be a copy of another manuscript originally compiled by Lady Ranelagh. Lastly, in the library of the Royal Society, there is a manuscript entitled *Medical Commonplace Book* used by Boyle and Lady Ranelagh. However, according to this historian, neither of these texts has been written by Lady Ranelagh herself [45].

Although Robert Boyle is best known today for his works on natural philosophy and experimentation, notably *The Sceptical Chymist*, his *Medicinal Experiments* (1692), published 1 year after his death, was one of his most popular books. Prior to its publication, he had already released a shortened version titled *Some Receipts of Medicine* in 1688. Indeed, in the 1680s, Boyle resumed writing texts of a medicinal nature, which had previously occupied him [46]. Although collecting recipes was a widespread hobby at the time, in Boyle's case, it was fueled by an inquisitive spirit and his own acknowledgment of having a fragile physical constitution [47]. As a result, in the later years of his life, Boyle meticulously made a selection of recipes for publication from his extensive collection of over a thousand accumulated formulations. These recipes demanded both chemical and metallurgical processes as well as galenic procedures. Several of them mention the assistance of his sister Katherine, both in the preparation and distribution of the resulting products. The significance Boyle placed on his recipes is evident in his last will, wherein he stipulated that all his manuscripts and collections of recipes, regardless of whether they were written by his own hand or others, were to be bequeathed to his sister, who actually passed away just a week before him [48].

4.3 Mad Madge

While Lady Ranelagh was a model of sobriety and conciliation, her contemporary Margaret Cavendish, Duchess of Newcastle (Fig. 4.1), was in every way the opposite. Margaret Lucas (1623–1673), born into the gentry, began serving as lady-in-waiting to Queen Henrietta Maria in 1643, who was the wife of Charles I and the mother of the future kings Charles II and James II of England. Next year, following the political upheavals of the nation, Margaret Cavendish accompanied the monarch in exile to Paris, where she met William Cavendish (1593–1676), an expatriate monarchist and widower, which ultimately lead to their marriage. Despite the 30-year age difference, the union with the future Duke of Newcastle was a happy and fruitful one. Rich, influential, and cultured—he had been tutored by none other than the philosopher Thomas Hobbes (1588–1679)—William Cavendish always supported his wife's intellectual pursuits and literary endeavors.

Thanks to her husband and his brother Charles, Margaret Cavendish had the opportunity of becoming acquainted with renowned philosophical thinkers throughout her time in Paris. Among them were notable French luminaries such as René Descartes, Pierre Gassendi, and Marin Mersenne, as well as influential English intellectuals, including Thomas Hobbes and Kenelm Digby, who were also exiled in the French capital [49]. Their discussions on mechanical philosophy and atomism have deeply impressed her [50]. In 1653, while in Antwerp (where she remained until returning to England in 1660), she began an intense period of writing that spanned various literary genres, including poetry, fiction, biography, theater, and essays on

Fig. 4.1 *Margaret Cavendish, Duchess of Newcastle* (1665), Peter Lely. Harley Gallery and Foundation

natural-philosophical themes. Her first work, *Poems and Fancies*, encompasses a vast collection of poems about atoms and the void. Regarding the former, she delves into their qualities (weight, size, shape) and explores how, through combination, they give rise to matter.

The atomist doctrine, of which the Cavendish Circle was an important herald, was slowly spreading across Europe. It all (re)started in 1417, when a copy of *De Rerum Natura* [*On the Nature of Things*] was discovered in a monastery in Germany. This poetic-philosophical work, transmitting an atomist vision of the world, was written in the first century BC by the Roman epicurean Titus Lucretius Carus (see Chap. 1) [51]. One of the first modern revivalists of atomism was the Italian monk Giordano Bruno (1548–1600), but its great disseminator was Pierre Gassendi (1592–1655), who, in his works *Animadversiones* [*Observations*, 1649] and *Syntagma Philosophicum* [*Philosophical Syntagma*, 1658], intended to demonstrate how Christianity and Epicureanism could harmonize. Although the Church considered the *De Rerum Natura* a dangerously heterodox work due to its rejection of religion, belief in the immortality of the soul, and its cosmological vision of infinite worlds, natural philosophers showed a growing interest in atomism [52, 53]. As a consequence, the new natural philosophy, the embryo of modern science, presented itself "visibly replete with atoms, corpuscles, and particles of all kinds" [54]. It was at this conjuncture that Margaret Cavendish built her own anti-Aristotelian natural philosophical system and published *Poems and Fancies*. In the poetic composition *A World made by Atoms* (whose first four lines open this chapter), she narrates the building of the natural world by atoms that, in their movements (not random, but in the form of a "dance"), find the place where they juxtapose with each other, like bricks or stones in the walls of a house.

The absence of randomness in the atomic dynamics envisioned by the Duchess of Newcastle suggests that she understood that atoms had their own order. Unlike Gassendi and other thinkers who aligned atomism with Christianity by attributing the movement of matter to God, the Duchess' atomistic vision did not incorporate the divine. Instead, her poems undeniably exhibit an Epicurean interpretation [55]. She may have been familiar with *De Rerum Natura*, but she could not have read it in Latin as it was a language she did not master, a reason that also deprived her of reading Gassendi's *Animadversiones*. On the other hand, the first English synthesis of this philosopher's work, due to Walter Charleton, a member of the Cavendish Circle, did not emerge before 1654—a year subsequent to the publication of *Poems and Fancies*. Furthermore, John Evelyn's English translation of Book I of Lucretius' work would only come about in 1656 [53, 56]. Nevertheless, due to her close association with Charleton, with whom she exchanged correspondence, and her discussions with her husband and brother-in-law, she would have been familiar with Lucretius' poem and the principles of atomism.

It is important to clarify that the theme of atoms in English literature was not novel in 1653. The resurgence of atomism had already furnished *material* for metaphors crafted by poets, authors of sermons, and political pamphleteers [57]. The depictions employed to render the atom *perceptible* to the human intellect possessed the ability to both comfort and evoke fear, often accompanied by a moral dimension. For example, to the poet John Donne (1572–1631), atomism presented itself as a doctrine of human essence that partially assuaged his materialistic concerns about death and resurrection [58]. His poem "An Anatomy of the World," written in 1611 as a eulogy for his patron's daughter, is an atomistic (and heretical) allegory of the Fall of Man and the end of the world [59]. The negative connotation of the divided, the disaggregated—by association with death and political, moral, and social disintegration—reveals the fear that the idea of the "atom" exerted on the popular imagination at the beginning of the seventeenth century. As historian Stephen Clucas has observed, "the *new philosophy*, which *puts everything in doubt*, which multiplies worlds in dizzying infinity, also dissects or atomizes the planet itself, *collapsed again into its atoms*" [57].

It did not take long until the Duchess of Newcastle started to harbor doubts about the likelihood of the "dance" of atoms she had extolled resulting in a stable universe governed by discernible laws. Furthermore, if it were indeed true that each atom possessed an equal power, as postulated by atomism, it would pose a threat to the hierarchical order deemed essential for the functioning and stability of systems, be they material or political. Such a doctrine had the potential to undermine the established structures, leading to the risk of descending into anarchy. Indeed, just 3 months after *Poems and Fancies* her second book, *Philosophical Fancies* (1653) came out. This prose work signified the inception of an anti-Aristotelian approach to interpreting Nature, presenting an alternative to the mechanistic perspective. Atomism, as a theory of matter, no longer played a role in this development, particularly in the original Epicurean form that she initially advocated. She would retain it, however, as a metaphor for politics and for the mind, both regarded as problematic atomistic systems [60]. Thus, in 1655, in *Philosophical and Physical Opinions*, she exposed for the first time a materialist-vitalist system according to which Nature was a body composed of several interconnected parts and that moved by itself. In *Observations upon Experimental Philosophy* (1666), during the peak of the critique against atomism, the Duchess of Newcastle asserted that "Nature is a perpetually-moving body, dividing, composing, changing, forming, and transforming her parts by self-corporeal figurative motions; and as she has infinite corporeal figurative motions, which are her parts, so she has an infinite wisdom to order and govern her infinite parts" [61]. To this, the Duchess of Newcastle added, "the opinion of atoms, is fitter for a poetical fancy, than for serious philosophy; and this is the reason that I have waived it in my philosophical works" [62]. Following her earlier works, she went on to write *Grounds of Natural Philosophy* (1668), which is regarded as the most succinct and well-structured presentation of her natural system. In this work, she posited that Nature consisted of three distinct types of matter, each infinitely divisible: rational animate matter, sensitive animate matter, and inanimate matter. The "parts" or constituent particles of nature, always containing the three types of matter, would possess sensory abilities, as well as reason and inherent motion [63].

While Margaret Cavendish departed from atomism as a doctrine of matter, numerous authors argue that she has retained a significant aspect of the atomic hypothesis within her materialist-vitalist natural system. This can be observed prominently in her elucidation of the qualities of matter, which she attributed to the forms of corporeal entities she referred to as "parts," "particles," and "figures," rather than using the term "atoms" [64, 65]. According to Clucas, contrary to what the title of the preface to *Philosophical and Physical Opinions* (1655) implies—"A Condemning Treatise of Atoms"—her position was not actually against atomism. Instead, it represented an intellectual refinement of the Epicurean-inspired atomism that she had previously expressed in *Poems and Fancies* 2 years earlier. The Duchess of Newcastle apparently held a stance opposing the concept of mechanistic atomism. She rejected the notion that the random collisions and movements of atoms, which she likened to "scattering like dust and ashes in the wind," could account for the structured composition observed in the material world [64]. She believed that the mobility of Nature's parts was not a result of their mechanical activation, as proposed in classical atomism. Instead, she attributed it to matter itself. In her view, matter's innate motion gave rise to its "figures," which in turn determined its material qualities. She considered motion, "figure," and matter as inseparable and intimately interconnected, with motion serving as the fundamental principle underlying it all. This aspect of her natural philosophy was intricately connected to the concept of "chemical atomism" of the German Daniel Sennert (1572–1637). Sennert regarded atoms not merely as material particles but as dynamic "units of action," which bears a certain resemblance to Thomas Hobbes' notions of the "natural motion" of atoms, as articulated in his work *De Corpore* (1655) [64]. Thus, aligning with the idea that the properties of substances are contingent upon the internal motions or "figures" of their parts, Margaret Cavendish developed a classification of "figures of motion" to account for various phenomena. For instance, she described fire as a material entity, with its characteristics arising from "expulsive" and "dilating motions" that give rise to pointed "figures" that "pierce, or shut, or wedge in sharp tooth" in the ignited matter, disassociating its parts. She delineated three types of fire, each characterized by a distinct kind of "burning figure." The first was the "bright-shining hot-burning fire" or flames, where the motions, "figure", and matter were all identical. The second type encompassed the "hot burning fire, but not a bright-shining fire," associated with corrosive substances like *aqua fortis* (nitric acid) or vitriol (sulfuric acid). The third type, referred to as "cold and dull," manifested in "fiery medicines" or piquant foods such as pepper. The first type of fire, with its unified motions, "figure," and matter, was formed by "lines of points which mounts upwards ... in a straight parallel line." The second type was composed of "sharp-edged lines like a razor or knife, or the like, which descends ... downward, or divide as ... streams of water, that digs itself a passage through the earth." The third type of fire was characterized by individual points [64, 66].

By 1668, Margaret Cavendish had already published 13 works, each featuring lavish editions with her name prominently displayed—an unprecedented accomplishment for a woman of her time. Fearlessly, she distributed them to renowned scholars at Cambridge and Oxford. In every aspect, she fostered a commitment to originality. Herself acknowledged her inclination toward singularity, even extending it to her choices in clothing and accessories [67]. The fact is that her eccentric clothes (designed by herself), her affected manners, and her intellectuality and advanced ideas (notably those related to women's education and animal rights) were a source of scorn for society. To a great number of people, she was considered insane, to the point where she was given the nickname "Mad Madge." Nevertheless, she remained utterly unperturbed by it all, and in fact, wherever she went, she caused a tremendous stir—she was a true socialite.

The exclusive Royal Society, consisting entirely of men, became the subject of her curiosity. One day, she expressed her desire to attend one of its sessions. Predictably, this caused a tremendous amount of embarrassment within the institution. However, due to the influential social and economic status held by the Dukes of Newcastle, a vote had to be taken by its members. Despite facing significant opposition, the majority voted in favor of her request. Thus, on May 30, 1667, the Duchess of Newcastle accomplished the remarkable feat of becoming the first woman to attend a session at the Royal Society. This unprecedented event would only be repeated much later in 1945 when crystallographer Kathleen Lonsdale (1903–1971) and biochemist Marjory Stephenson (1885–1948) were elected as full members, as will be discussed in Chap. 6 of next volume [68].

On that particular day, a session was meticulously prepared to bring delight to the Duchess, with Robert Boyle leading the scientific demonstrations while receiving assistance from Robert Hooke. The demonstrations comprised various experiments, such as chemical reactions that resulted in changes of color, the dissolution of mutton meat in sulfuric acid, the effect of magnetite on a compass placed two meters away, and the workings of the air pump. The Duchess, who, due to her tardy arrival with a delegation dispersed among numerous carriages, had missed the first part dedicated to society's common affairs, was thoroughly enthralled by the experiences. Her response was one of sheer admiration, leaving her virtually speechless [69, 70].

Her desire to attend a session of the Royal Society was, however, surprising, especially because the previous year, in *Observations upon Experimental Philosophy*, she had criticized most of the experimentalist members of that society, and in the same work she also rejected the vacuum and the use of instruments of observation with lenses, such as the telescope and the microscope. (Robert Hook, in his *Micrographia* (1665), had explained the advantages of their use.) Regarding these optical devices, about which ironically the Duchess said were "currently so applauded," she stated that they were "artificial instruments" that could prove to be "deluders, rather than true informers" [71]. Later, in *Grounds of Natural Philosophy*, she would again criticize the followers of experimentalism and the use of optical instruments, saying that "they waste their time and estates with fire and furnace, cruelly torturing the productions of Nature to make experiments," adding that they troubled themselves examining and looking through "telescopes, microscopes, and the like toyish arts, which neither get profit, nor improve their understandings" [72]. Samuel Pepys, member of parliament and future president of the Royal Society, attended the May 30 session and recorded in his famous diary what he saw and felt about her:

> The Duchesse has been a good, comely woman, but her dress so antique, and her deportment so ordinary, that I do not like her at all. Nor did I hear her say anything that was worth hearing, except that she was full of admiration, all admiration [73]

The Duchess of Newcastle's ostentation and pomposity, and especially her irreverence with regard to social and religious conventions, made her a most disliked figure in the eyes of the discreet and pious Lady Ranelagh [74]. The latter did not hesitate to correspond with her brother, Richard, Earl of Burlington, regarding the acceptance of the Duchess into the Royal Society. In her letter, she remarked, "I am resolved she escapes Bedlam only by being too rich to be sent there, but she is mad enough to convey the title of her residence to that of an asylum" [75]. As early as 1653, when *Poems and Fancies* were published, another lady named Dorothy Osborne, who would later become Lady Temple, had also described the Duchess using similar terms. In a letter to her then-fiancé, statesman, and essayist William Temple, she criticized the Duchess's boldness in the following terms: "Surely the poor woman is a little distracted; she could never be so ridiculous else as to venture at writing books, and in verse too." In a subsequent letter, she informed him that she had already laid eyes on the book, remarking that "there are many soberer People in Bedlam" [76].

In 1952, Samuel Mintz was the first historian to make a detailed analysis of Margaret Cavendish's visit to the Royal Society. He painted a similarly devastating picture of her, helping to feed the idea that she was an anti-experimentalist who had been unable to understand the scope of the methods employed by figures such as Boyle or Hooke. Observing that the Duchess, with her forays into natural philosophy, was seeking nothing but pleasure, Mintz recalls the writer Virginia Woolf, who said of her that she was always "ecstatic in thought" [69]. Indeed, the British writer, in the seventh of her essays in *The Common Reader* (1925), had referred to the Duchess of Newcastle in an unsavory tone, stating that she had "the irresponsibility of a child and the arrogance of a duchess" and that there was "something noble and Quixotic and high-spirited, as well as crack-brained and bird-witted, about her" [77].

For many centuries, Margaret Cavendish's literary and philosophical works were dismissed or disregarded as irrelevant and lacking coherence. However, in recent decades, there has been a growing interest in her contributions to the natural-philosophical discourse of the time. This renewed attention can be attributed, in large part, to the efforts of feminist and women's history scholars who have shed light on her significance [78]. Four decades ago, the renowned ecofeminist historian and philosopher Carolyn Merchant described her as "[a] feminist who between 1653 and 1671 wrote some fourteen scientific books about atoms, matter and motion, butterflies, fleas, magnifying glasses, distant worlds, and infinity" adding that "her ideas and theories are often inconsistent, contradictory, and eclectic, which is attributable at least in part to her lack of formal education—a lack she herself deplored [79]."

Indeed, Margaret Cavendish's forceful denunciation of the precarious state of women's education in her time, including her own, as pointed out by Merchant, has played a significant role in her frequent recognition as a proto-feminist [80]. An exemplary instance of this can be found in her work *Sociable Letters* (1664), where she directly tackles the topic of aristocratic women's education and their disposition toward philosophical-natural knowledge:

> [...] for the most part women are not educated as they should be, I mean those of quality, for their education is only to dance, sing, and fiddle, to write complemental letters, to read romances, to speak some language that is not their native, which education, is an education of the body, and not of the mind [81].

[…] neither does our sex delight or understand philosophy, for as for natural philosophy they study no more of Nature's work than their faces, and their greatest ingenuity is, to make them fairer than Nature did […] [82]

Following the current trend of reevaluating Margaret Cavendish's intellectual significance, historian Emma Wilkins provides a nuanced perspective on her relation with the Royal Society. Wilkins acknowledges that despite certain reservations about experimentalism, the Duchess of Newcastle actively endorsed crucial elements of the institution's agenda. Specifically, she supported the use of everyday language to disseminate novel concepts in natural philosophy and emphasized the importance of an open discourse free from dogmatism. Another factor that drew Margaret Cavendish closer to the Royal Society was her commitment to aligning natural philosophy with the betterment of human welfare, a principle vigorously advocated by Francis Bacon. In this regard, she specifically commended Robert Boyle for his clear and accessible language, noting that he did not complicate Nature or obscure truth with extravagant terminology or complex language [83]. By her turn, Sarah Hutton highlights the Duchess of Newcastle's direct contact with prominent European intellectuals of her era, emphasizing that interpreting Cavendish's natural philosophy within the framework of works like Thomas Hobbes' *De Corpore* reveals her ideas to be more than a mere reflection of the intellectual climate of the time. Instead, they represent a valuable and original contribution [84]. In fact, Margaret Cavendish addressed, among other topics, Hobbes' materialism (which she rejected), Cartesian dualism, and mechanistic views on Nature and causality [85]. Her natural philosophy, despite materialistic, conveyed a vision that embraced organic and vitalist principles [84]. In regard to her disdain for scientific instruments utilizing lenses, leading to assertions such as "the observation of a bee through the microscope does not bring [the observer] more honey" [86], it is true that she was familiar with them. During the exile in Paris, her husband obtained several microscopes and telescopes. Consequently, she was aware that, while not yet fully effective, these instruments only offered insights into the external characteristics of the analyzed objects [87]. Her skepticism toward these scientific observation devices finds some parallel in the essay on microscopic anatomy authored by philosopher John Locke and physician Thomas Sydenham 2 years later [88]. Moreover, Margaret Cavendish's perspective on observation through instruments was part of a larger debate on the respective roles of art (in this context, instruments) and Nature in scientific inquiry [87].

Authors like Deborah Boyle challenge the tendency to view Margaret Cavendish's vitalist materialism, along with her rejection of mechanical philosophy and atomism, as indications of protofeminism. They argue that analyzing her natural philosophy in terms of its intrinsic arguments does not diminish its value or historical significance [89]. While Lisa Sarasohn characterizes her work as "largely uncritical and hopelessly repetitive," she also acknowledges that Cavendish's natural philosophy is no more fantastical than that of certain male contemporaries. Notably, figures like Kenelm Digby and Johannes van Helmont, who were esteemed as scientific prodigies and pioneers in their time, provide examples of male counterparts whose ideas were similarly visionary [90]. Digby, in spite of his attempt to reconcile Cartesianism and atomism with Aristotelianism—resulting in a hybrid devoid of any systematic approach—and his dedicated pursuit of medicinal cures through paranormal practices, stood as one of the founding members of the Royal Society. On the other hand, Van Helmont, while an experimenter, also delved into hermeticism, for during the seventeenth century, natural philosophy (science) and magic (pseudoscience) remained closely intertwined. In our days, some consider that the metaphysical notion of Nature as universally rational and sentient indeed stands as one of the Duchess of Newcastle's most original contributions to modern natural philosophy. She was not, moreover, the sole proponent of this viewpoint; the Dutch philosopher Baruch de Spinoza (1632–1677) and the German polymath Gottfried Wilhelm Leibniz (1646–1716) shared a similar understanding of the world as fundamentally perceptive [91].

One of Margaret Cavendish's most captivating works is *The Blazing World* (1666) [92]. This novel, regarded as a pioneer of science fiction, unfolds in a utopian kingdom situated in another realm, inhabited by peculiar beings who lack a true understanding of science and philosophy. The main character is a young woman who ascends to the role of empress and becomes the educator of such creatures. There have been suggestions that the inspiration for this character was Christina of Sweden, whom the Duchess of Newcastle personally met in Antwerp in 1655. Similar to the Swedish monarch, the empress in *The Blazing World* is portrayed as an esoteric alchemist and atomist, establishing an academy for humanists and scientists [93].

4.4   The Restless Queen

In the course of the seventeenth century, chemistry experienced a process of divergence from its alchemical origins. Inspired by iatrochemists, certain experimentalists discovered previously unknown chemical reactions and embraced new ideas aimed at understanding the nature of matter, rather than pursuing the Philosopher's Stone. While much of the chemistry in the 1600s remained rooted in Paracelsus' triad, the concept of the atom slowly began to resurface [94].

One notable practitioner of Paracelsian alchemy in the late seventeenth century was Christina of Sweden (1626–1689), a monarch known for her unconventional interests and her abdication from the throne in 1654, which she had inherited at the age of six following her father Gustav Adolf II's death in battle. Another extraordinary facet of her life was her conversion to Catholicism in 1655 and her relocation to Rome. She made a triumphant entry into the city through the Piazza del Popolo and resided there until her passing. In the Accademia Reale, initially housed in Palazzo Farnese and later in Palazzo Riario (now Palazzo Corsini), she surrounded herself with figures from the realms of culture and science, including alchemists and esoteric thinkers.

Christina of Sweden believed in an *anima mundi* (world soul), a cosmological concept derived from Plato, according to which there would be a universal spirit or force underlying all of Nature that endowed it with form and movement, just as the human soul animated man. This same idea is evident in several Eastern doctrines and similar concepts having been proposed by figures such as Paracelsus, Baruch de Spinoza, or Gottfried Wilhelm Leibniz (as discussed above). According to the latter, who visited the monarch in Rome in 1689, her idea of the *anima mundi* was, however, more that of the Arab Averroes (c.1126–1198) than that of Plato.

In 1655, in France, when asked about her true religion, Christina of Sweden replied that it was philosophy, which, although "indeterminate and with uncertain limits, was nowhere so well represented as in Lucretius' *De Rerum Natura*" [95]. Lucretius had described a world made up of atoms, which incessantly aggregated and disaggregated in an infinite and random number of sequences. In such a world, the power of faith or divine order did not enter. Despite the classic Lucretian atomism, the monarch, who had a copy of the treatise *Animadversiones*, was also influenced by the contemporary vision of Pierre Gassendi. In his work, the philosopher argued that if the number of existing atoms were not infinite, but rather limited, God could exercise His guardianship of the world and, thus, control the way in which those were organized. The sovereign reportedly invited Gassendi to visit her in Stockholm, although this meeting never took place [96]. The French philosopher and mathematician, who attempted to Christianize atomism so that the combination and dissociation of atoms would not be the result of mere chance, had been recommended to Christina of Sweden by her personal physician, Pierre

Bourdelot (1610–1685). Charismatic and open-minded, a true freethinker with an atomistic conception of the universe, Bourdelot, despite being an abbot, was also an atheist and likely influenced the monarch. To help her overcome her chronic melancholy, he is said to have advised her to put an end to her obsessive reading habits (while, at the same time, providing her with literature that he deemed suitable).

Descartes was also invited to the Swedish court. The renowned philosopher accepted (Fig. 4.2), albeit with tragic consequences. He arrived in Stockholm in early October 1649, but the philosophical and religious lessons desired by the monarch did not commence until just before Christmas. Adding to the difficulty, these sessions were scheduled for five o'clock in the morning, in a cold and drafty castle. In addition to the lack of strong rapport between the two, Descartes was not accustomed to rising early, and he took great pleasure in having well-heated rooms where he could comfortably meditate (as mentioned at the beginning of this chapter). He fell seriously ill on February 1, 1650, and succumbed to

Fig. 4.2 *Cristina of Sweden discussing with René Descartes* (1884), Nils Forsberg after Pierre-Louis Dumesnil the Younger. National Museum of the Palaces of Versailles and Trianon

pneumonia 10 days later. Some alleged that he had been intentionally poisoned to hinder his potential conversion of the Lutheran monarch to Catholicism [97].

Christina of Sweden's library, now a part of the Vatican Apostolic Library, consisted of 4500 printed books and 2200 manuscripts, including numerous alchemical works spanning from ancient to contemporary times. Additionally, within her palace in Rome, she maintained a laboratory where various prominent figures in the field of alchemy would frequently visit. In her writings on this discipline, the monarch expressed her thoughts:

> [...] it is a beautiful science. It is the anatomy of Nature and the true key to all treasures. It brings fortune, health, glory and true wisdom to its holder [93]

There is no precise information regarding the exact date when Christina of Sweden began her study of alchemy. However, it is possible that her interest was sparked by Johannes Franck (1590–1661), a Swedish physician, botanist, and alchemist who served as the rector of the University of Uppsala. In his work titled *Colloquium Philosophicum* Cum *Diis Montanis* [*A Philosophical Conversation with the Gods of the Mountains*, 1651], Franck encouraged Christina to pursue the search for the Philosopher's Stone [98]. He believed that her reign embodied the fulfillment of Paracelsus's prophecy of the return of Elias the Artist. According to the Swiss alchemist, Elias would bring forth the revelation of many secrets, including the transmutation of metals, which would lead to a renewal of the arts and sciences.

The monarch's fascination with Paracelsian alchemy was rooted in the prevailing Hermetic-Christian currents of thought in the latter half of the seventeenth century. It brought her into contact with Rosicrucianism and Kabbalah, as evidenced by the contents of her extensive library. Over time, she became increasingly devoted to alchemical research and received assistance from a young woman named Sibbyla. Sometimes, she conducted her experiments in the laboratory alongside notable visitors such as the renowned Italian physician, esotericist, and adventurer Giuseppe Francesco Borri (1627–1695), as well as the German Jesuit and polymath Athanasius Kircher (1602–1680). Among her correspondents was Johan Rudolf Glauber (1604–1670), the discoverer of the "mirabile salt," also known as "Glauber's salt." This compound, sodium sulfate decahydrate ($Na_2SO_4.10H_2O$) [99], held special significance for him as he viewed it as the manifestation of Elias the Artist's prophetic return.

The Marquis Massimiliano de Palombara (1614–1680), a member of the sovereign's circle in Rome, had the "Door of Alchemy" (also referred to as the "Hermetic Door" or "Magic Door") erected in the garden of his palace on the

Esquiline Hill. This gate, currently situated in Piazza Vittorio Emanuele II, features inscriptions and alchemical symbols. It was intended to commemorate an alleged successful transmutation that took place in the sovereign's laboratory. However, there is a scarcity of available records regarding her alchemical endeavors. A drawing of laboratory vessels, accompanied by brief notes on calcination and heating times, is known to have been executed by Christina of Sweden herself. Furthermore, there is a document in her handwriting titled *Il Laboratório Filosofico, Paradossi Chimici* [*The Philosophical Laboratory, Chemical Paradoxes*], which could be either an outline for a presentation of hers on alchemy or comments on a text with a similar title. Historian Susanna Åkerman suggests that the paradoxes referred to in the document's title may have been influenced by Robert Boyle, whose work *The Sceptical Chymist* Christina owned in Latin translation. Boyle's original subtitle for the work is *Chymico-Physical Doubts & Paradoxes* [93].

4.5   The Mind Has No Sex

Descartes did not specifically address the female sex, but the universality of his mind-body dualism necessarily placed him at the center of discussions about the role and value of women. The customary theories of female physical and mental inferiority, based on the comparison of a woman's body with that of a man, could now be refuted with modern arguments. However, there is no consensus on whether the great French philosopher ever considered the possibility of the new natural philosophy being receptive to female involvement. If he had entertained such hopes, they would have been in vain since, from the very first moment in 1666, the Académie Royale des Sciences barred entry to the *second sex*. In any case, the new philosophical-scientific knowledge aroused great interest among female audiences, both aristocratic and middle class. Descartes himself benefited from two female patronages—in addition to that of Christina of Sweden, he also had that of Princess Elisabeth of Bohemia [100, 101].

   In France, from the beginning of the seventeenth century, salons evolved into genuine temples of learning and cultural display for privileged women. Every week, *salonnières* opened their homes to host members of society and intellectual figures who engaged in discussions about new literary works, scientific advancements of the time, and the latest political events. The salon was a creation of Catherine de Vivonne, Madame de Rambouillet (1588–1665), who, due to her poor health, could not attend the court. To overcome this, she established a miniature private court within her Parisian residence, consisting of aristocratic, cultured, and fashionable individuals. Within these salons, gallant conversations took place, with ideas and witty remarks exchanged, always following the etiquette rules defined by the hostess. Madame de Rambouillet, finding the manners of Louis XIII's court too crude, drew inspiration from the ancient codes of chivalry, particularly influenced by Castiglione and his work *Il Libro del Cortegiano* (see previous chapter) [102]. One label forever associated with the *salonnières* of this century was *précieuses*, meaning "precious" in the pedantic sense. Madeleine de Scudéry (1607–1701), another renowned *salonnière* and successful writer, despite advocating for female education discussed learned women in her book *Les Femmes Illustres* [*The Illustrious Women*, 1642], comparing them to those who exhibited true pretentiousness. Molière, known for targeting the bourgeoisie in his satires, used Scudéry as a model for the character Magdelon in his comedy *Les Précieuses Ridicules* [*The Precious Ridicules*, 1659], where he mocked women who sought to refine conversation and the French language [103, 104]. Another pejorative term applied to women was *savantes*, meaning "know-it-alls," referring to those who liked to show off their knowledge. This brings to mind the old Portuguese saying, "From the she-donkey that brays and the woman who knows Latin, save yourself and me!." In another of Molière's comedies, *Les Femmes Savantes* [*The Learned Ladies*], first performed in Paris in 1672, the female characters who parroted cultural themes and struggled with the names of philosophers were contrasted with the modest woman who concealed her knowledge. Clitandre, the young man in love with Henriette, expressed the ideal of an enlightened yet discreet woman:

I consent to a woman knowing everything,
But I do not admit in her the shocking passion
Of acquiring knowledge to make herself wise;
And I like her to, when questions are asked,
Often feign ignorance of things she knows;
I want her to hide her studies,
And possess knowledge without seeking recognition,
Without quoting authors, without using grand words,
And demonstrate subtlety in her few words [105]

It was within this sociocultural backdrop that the publication of what is considered the first book on chemistry written by women took place in Paris in 1666. The book, titled *La Chymie Charitable et Facile, en Faveur des Dames* [*Charitable and Easy Chemistry for the Benefit of Ladies*], was authored by Marie Meurdrac (1610–1680). The initial Paris edition was followed by two more editions, with numerous reprints until 1711. Additionally, there was an edition published in Lyon in 1680, six editions in Germany between 1673 and 1738, and one edition in Italy in 1682 [106]. Meurdrac's work measures 8.5 × 15 cm and contains 334 pages. Additionally, there are 36 pages comprising a dedication, six poems, a certificate of approval from the Paris Faculty of Medicine, the royal authorization for printing, an index, and a preamble [107]. Pages 39 to 42 list 106 alchemical symbols, while page 43 presents a table of weights used in medicine (this information is based on the second edition printed in Paris in 1674). From the third French edition on, it included a frontispiece accompanied by an illustration depicting a female figure raising a curtain and gesturing toward a series of containers, flasks, and laboratory funnels, symbolically promising to reveal the secret world of chemistry to its readers (Fig. 4.3).

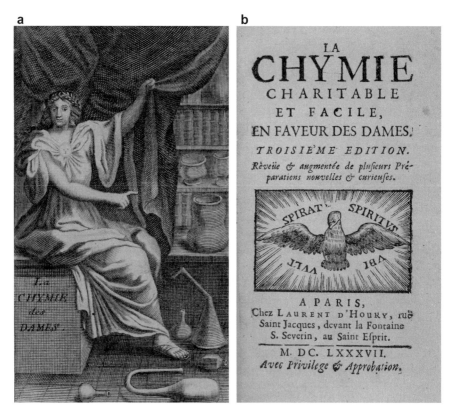

Fig. 4.3 Illustration and frontispiece of the third edition of *La Chymie Charitable et Facile, en Faveur des Dames* (Paris, 1687). Courtesy of the Science History Institute

Regarding the author, it is known that she came from a family with aristocratic lineage on her father's side. She spent her early years in the family mansion located in the village of Mandres, on the outskirts of Paris. After marrying Henri de Vibrac in 1625, she moved to the palace of Grosbois, owned by the Duke of Angoulême, where her husband held the position of commander of the guard unit. It was during this period that she had the opportunity to meet the Countess of Guiche, whom she served as a lady-in-waiting and formed a close friendship with (and to whom she dedicated her book). A significant portion of Marie Meurdrac's life was dedicated to religion and pharmaceutical chemistry. She conducted her investigations in her own laboratory, creating ointments and preparations that were distributed by both herself and the Countess of Guiche to poor people in the region. This allowed her to witness the effectiveness of her creations [108].

In the preface of her work, Meurdrac acknowledges that she is fully aware of being the first woman to publish a book on chemistry. She explains that her decision to publish her "little treatise" was driven by her personal satisfaction and the desire to preserve the knowledge she had acquired through extensive and repeatedly confirmed experimentation. However, she admits that this decision was not an easy one for her to make: "it was not a woman's task to teach; it was up to her to remain silent, to listen, and to learn without testifying her knowledge," all the more so because "men always despise and denounce the productions coming from a woman's thought." On the other hand, she adds that she is not the first woman to publish something and that, moreover, "minds have no sex" and argues that "if women's minds were cultivated like men's, and if equal time and expense were invested in their instruction, they could equal them," noting that "the century has seen the birth of women who in prose, poetry, languages, philosophy, and even the government of the State, owe nothing to the competence and ability of men." In support of her decision to publish, Meurdrac emphasizes the usefulness of her work: "it contains many infallible remedies to cure illnesses and to preserve health, as well as several rare secrets ladies; not only to preserve them but also to increase the benefits they have received from Nature." She still adds: "It would be a sin against charity to hide the knowledge that God has given me, which can help the whole world. This is the only reason why I let this book out of my hands." At the end of the *avant propos*, this is how Meurdrac describes her work:

> I have divided this book into six parts: the first deals with principles and operations, vessels, seals, furnaces, fires, symbols and weights; in the second I speak of

the virtues of the simple remedies, their preparations and the method of extracting their salts, tinctures, waters and essences; the third is about animals; the
fourth treats of metals; the fifth is about the way of making the compound
remedies, with several medicines all tried; and in the sixth, for the benefit of the
ladies, I speak of all things capable of preserving and enhancing beauty. I did
everything I could to explain myself well and facilitate the operations; I have
never wished to go beyond my knowledge, and I can assure you that everything
I teach is true, and that all my remedies have been tried; for which I praise and
give glory to God

At the very beginning of the first chapter, she informs that:

[...] chemistry has for its object mixed bodies, insofar as they are divisible and
soluble, on which it operates by extracting from them the three principles, salt,
sulfur and mercury, which is done by two general operations, namely, dissolution and freezing

Next, on page 8, she states her opinion regarding the advantages of practice
over theoretical explanations:

Those who have written on these matters are less attached to demonstration
than to speculation, so that they are often mistaken, because theory and practice
are generally different, and action teaches us more than contemplation

On page 58, in the paragraph discussing the medical use of rosemary
essence, we find a highly descriptive depiction of the nature of the work:

A modern philosopher was right in saying that any man who cherishes life
should keep rosemary essence at home, as an antidote to all kinds of ills. I myself
have successfully used it, achieving admirable cures. One can take six to ten
drops of its essence in wine or in a teaspoon of sugared water, in the morning on
an empty stomach. It serves as protection against any kind of infected air, treats
jaundice, dispels bad moods, uplifts the heart, alleviates melancholy, purges the
spleen, provides immediate relief for conditions such as apoplexy, tonsillitis and
lethargy. It freshens the breath, bestows a rosy complexion, sooths the stomach,
and aids digestion when taken as described

In Meurdrac's book, similar to other practical works dedicated to women
(discussed in the previous chapter), numerous recipes for cosmetic preparations and "rare secrets" for ladies are featured. These include the likes of the
"water against wrinkles," the "powder for teeth," the "ointment to remove

flour that appears at the roots of the hair" (dandruff), and the renowned "Queen of Hungary's Water." Regarding the latter, the author states that it is precisely the recipe left behind by the Hungarian monarch in her own hand-writing [109]. The preparation of this water, which according to Meurdrac was responsible for the queen's youthful appearance even at the age of 72, required two pounds of brandy distilled four times, to which 22 ounces of rosemary flowers and spikes were added. The mixture was left to steep for 50 hours in a tightly sealed container. Afterward, it was distilled using a water bath. The resulting preparation was intended to be taken once a week in the morning, in an amount equivalent to the weight of one drachma diluted in broth. Additionally, it was recommended for daily use in washing the face and could be applied to affected areas of the body. On page 13, Meurdrac provides a description of the water-bath distillation process, explaining that its name originates from its inventor, "the sister of Moses, Maria, also known as the Prophetess, who authored the book titled *Of the Three Words*."

In truth, despite the author's claims regarding her involvement in chemistry and the value she places on experimentation, the inclusion of the Paracelsian *tria prima* (salt, sulfur, and mercury—as discussed in Chap. 2) clearly reveals the enduring alchemical influence in her work. Supporters of the chemical aspects of *La Chymie Charitable et Facile* highlight the scientific nature of certain observations made by the author. For instance, she warns against the use of mercury compounds in cosmetics and emphasizes the challenges involved in preparing "oil of talc." This is unsurprising, given that talc is hydrated magnesium silicate, a mineral from which no oil can be obtained. Additionally, it has been noted that some of the preparations presented in Meurdrac's book can also be found in significant works on chemistry by male authors of the seventeenth century, such as Jean Béguin, Nicolas le Fèvre, Christopher Glaser, and Nicolas Lémery. Examples include the recipe for "Flowers of Benzoin" (benzoic acid crystals) and "Saturn's salt" (lead(II) acetate). In short, Meurdrac's book is generally regarded as an unpretentious representation of the therapeutic chemistry of the time (meaning iatrochemistry) while simultaneously serving as a testament to the emerging spirit of scientific inquiry [110, 111]. However, some simply categorize it within the tradition of "cookery medicine" books, as Meurdrac's understanding of chemistry seems to revolve primarily around distillation and the mixing of substances [112].

Based on the limited information available about Marie de Meurdrac, it is not unreasonable to speculate that she might have fit the mold of the knowledgeable woman depicted by Molière in his play *Les Femmes Savantes*, which was staged just 6 years after the publication of *La Chymie Charitable et Facile*

[113]. While this comedy contains references to fields such as philosophy, mathematics, astronomy, and physics, chemistry is not mentioned. This omission is not surprising, as this discipline had not yet gained universal recognition as a science, despite already being in vogue. A proof of this is the fact that Bernard de Fontenelle, the secretary of the Académie des Sciences (see Chap. 6), when praising the public chemistry courses that the pharmacist Nicolas Lémery (1645–1715) had begun on Rue Galande in Paris, in the very same year that Molière's play premiered, observed: "even women, attracted by the trend, dared to attend these erudite gatherings" [114]. Indeed, the public chemistry courses of Nicolas Lémery (who was one of the pioneers in the interpretation of acid-base reactions) reached high levels of popularity. In 1675, he published *Cours de Chymie* [*Course in Chemistry*], a work that was hailed as the unveiling of a completely new science "at last stripped of all references to occult qualities, of all barbaric and dark jargon" [115].

Another book of alchemical recipes for medical applications that appeared in seventeenth-century France was *Les Remèdes Charitables de Madame Fouquet* [*The Charitable Remedies of Madame Fouquet*, 1681]. It was authored by Marie de Maupeou (1590–1681), an aristocratic woman married to François Fouquet, Viscount de Vaux, a member of the Paris Parliament and a close associate of Louis XIII and Cardinal de Richelieu. One of her sons, Nicolas Fouquet, would later become the finance minister under Louis XIV. As a devout and compassionate Catholic, Madame Fouquet, like Marie Meurdrac, dedicated herself to assisting the impoverished by providing them with the remedies she prepared. Her book, which went through more than 50 editions in France, continued to be published well into the eighteenth century and received various versions and translations, including Italian, Spanish, German, and Portuguese. The Portuguese translation, titled *Recolha de Remédios Escolhidos de Madame Fouquet* [*Collection of Selected Remedies by Madame Fouquet*], was published in Lisbon in two volumes. The first one, presenting parts I and II as in the French editions, was published in 1712, while the second volume, presenting part III, came out in 1749. Prior to this, another of her works, *Recueil de Recettes Choisis, Expérimentées et Approuvées Contre Quantité de Maux Fort Communs* [*Collection of Selected, Experimented and Approved Recipes Against a Variety of Common Illnesses*, 1675], had been published at the request of her son Louis Fouquet, who was the Bishop of Agde [116, 117].

Upon analyzing Madame Fouquet's recipes, it becomes evident that they closely resemble the formulations found in official pharmacopoeias of the time. One such example is the recipe for the *Manus Dei plaster*, which, as its name suggests, acted as if it were the hand of God, healing an endless number of sores, ulcers, cuts, tumors, bites, stings, fistulas, burns, and more:

Take galbanum	1 ounce, 2 drachms
Ammonium gum	3 ounces, 3 drachms
Opoponax	1 ounce
White very strong vinegar	4 ounces
Olive oil	2 pounds & a half 1 pound
Gold litharge	1 pound, 1 ounce
Verdigris	1 ounce
New wax	20 ounces
Magnetite	2 ounces
Mastic	1 ounce
Olibanum	2 ounces
Bdellium	2 ounces
Myrrh	1 ounce, 2 drachmas
Frankincense	1 ounce, 2 drachmas
Aristolochia rotunda	1 ounce [118]

According to Lémery's *Universal Pharmacopoeia* (1697), that plaster was prepared as follows:

Take gold litharge, two pounds; common oil, four pounds; common water, three pounds. Boil them to the consistency of a plaster, then mix in yellow wax, one pound; turpentine of Venice, half a pound; ammonia gum, galbanum, opopanax, sagapene, myrrh, frankincense, mastic, four ounces of each. Laurel oil, three ounces; magnetite and calamine; *Aristolochia rotunda* and long, two ounces of each. Make a plaster according to the art [117]

In England, the "cookery medicine" of Joanna Stephens was an extraordinary case. This English woman, who mostly lived in the first half of the eighteenth century, but of whom very little is known, came from a respectable family from the county of Berkshire and practiced medicine for many years, likely operating a small hospital in London [119, 120]. Around the late 1730s, she devised several formulations for dissolving bladder stones, providing an alternative to traditional surgery. The therapeutic preparations she developed (for oral administration or direct injection into the bladder) consisted of pill, decoction, and powder forms [121]. The powder primarily comprised calcinated eggshells, while the decoction involved soap and various plants. The pills contained burnt burdock snails, seeds, and wild carrot, along with soap and honey [122].

Despite facing many accusations of quackery, including claims that she merely used ingredients known since antiquity to be useless or even dangerous, Stephens had her share of supporters. In 1738, David Hartley, an English philosopher and member of the Royal Society, asserted to have experienced great success with her treatment. He published a pamphlet titled "Ten Cases of People Who Have Taken Mrs. Stephens' Stone Remedy," expressing his gratitude and even proposing to raise a sum of five thousand pounds to reward the creator of such an extraordinary remedy [123]. Hartley initially attempted a private subscription but received only a small amount, so he presented a petition to parliament. To support his case, he documented 155 successful cases of the remedy's effectiveness. In June 1739, Stephens submitted her prescription to the Archbishop of Canterbury, and it was promptly published in *The London Gazette* [120]. The preparation underwent chemical analysis by English and French scientists, including members of the Royal Society and the Académie Royale des Sciences. The treatment's efficacy was monitored in four patients before and after the intervention. Finally, in March 1740, parliament ruled in favor of Hartley's proposal, and Joanna Stephens received her reward. She disappeared from public view thereafter, but contrary to earlier predictions, the concept of dissolving or fragmenting calculi persisted and continued to be developed. By the end of the nineteenth century, hundreds of treatises on the subject had been published. In the nineteenth century, Alexander Marcet (husband of Jane Marcet, one of the heroines of this book—see Chap. 6), a physician and analytical chemist renowned for his expertise in gallstones and urinary tract stones, authored *An Essay on the Chemical History and Medical Treatment of Calculous Disorders* in 1817.

References and Notes

1. Margaret Cavendish, *Poems, Or, Several Fancies in Verse: With the Animal Parliament in Prose,* London, 1668, p. 7
2. Theological and philosophical concept that emerged in the Middle Ages, according to which Nature was a book that had to be read in order to be understood.
3. Coordinate system that allows representing each point in a plane defined by two perpendicular lines, using a pair of numerical coordinates that correspond to the distances from the point in question to the origin, which is the point where the two lines intersect (point 0,0).
4. Patricia Fara, *Science—A Four Thousand Year History*, Oxford University Press, Oxford, 2009, pp. 125–130
5. Margaret Wertheim, *Pythagoras's Trousers—God, Physics, and the Gender Wars*, Fourth Estate, London, 1997, pp. 94–97
6. Bertrand Russel, *História da Filosofia Ocidental Vol. 2*, Círculo de Leitores, Lisboa, 1977, pp. 123–129

7. Descartes' Physics, *in Stanford Encyclopedia of Philosophy*; https://plato.stanford.edu/entries/descartes-physics/#SpacBodyMoti

8. C. Meinel, "Early Seventeenth-Century Atomism: Theory, Epistemology, and the Insufficiency of Experiment", *Isis*, 79 (1988) 68–103

9. M. E. Ehrlich, "Mechanism and activity in the scientific revolution: The case of Robert Hooke", *Annals of Science*, 52 (1995), 127–151

10. Wertheim, *Op. cit.* (5), pp. 97–98

11. M. Biagioli, "Knowledge, Freedom, and Brotherly Love: Homosociality and the Accademia dei Lincei", *Configurations*, 3 (1995) 139–166

12. John Gribbin, *Science: A History*, Penguin, London, 2003, p. 159

13. Wertheim, *Op. cit.* (5), pp. 99–100

14. Londa Schiebinger, *The Mind Has No Sex—Women in the Origins of Modern Science*, Harvard University Press, Cambridge (MA), 1989, p. 26

15. J. E. Archer, "Women and Chymistry in Early Modern England: The Manuscript Receipt Book (c. 1616) of Sarah Wigges", *in* Kathleen P. Long, (ed.), *Gender and Scientific Discourse in Early Modern Culture*, Routledge, Abingdon, 2010, p. 197

16. P. Bayer, "Lady Margaret Clifford's Alchemical Receipt Book and the John Dee Circle", *Ambix*, 52 (2005) 274–284

17. P. Bayer, "Women Alchemists and the Paracelsian Context in France and England, 1560–1616", *Early Modern Women*, 15 (2021) 103–112

18. Quoted in reference 20

19. E. Bourke, ""I would not have taken her for his sister": financial hardship and women's reputations in the Hartlib circle (1641-1661)", *The Seventeenth Century*, 37 (2022) 47–64

20. M. Rayner-Canham, G. Rayner-Canham, "British women and chemistry from the 16th to the mid-19th century", *Bull. Hist. Chem.*, 34 (2009) 117–123

21. Marelene Rayner-Canham, Geoffrey Rayner-Canham, *Pioneering British Women Chemists: Their Lives and Contributions*, World Scientific, London, 2020, p. 3

22. M. Boas, "Quelques aspects sociaux de la chimie au XVII e siècle", *Revue d'histoire des sciences et de leurs applications*, 10 (1957) 132–147

23. Michelle M. DiMeo, *Katherine Jones, Lady Ranelagh (1615-91): Science and Medicine in a Seventeenth-Century Englishwoman's Writing*, PhD Thesis, University of Warwick, 2009, pp. 1–6

24. M. DiMeo, "Openess *vs.* Secrecy in the Hartlib Circle: Revisiting 'Democratic Baconianism' in Interregnum England", *in* Elaine Leong, Alisha Rankin (eds.), *Secrets and Knowledge in Medicine and Science, 1500–1800*, Routledge, London, 2011, p. 105

25. E. Bourke, "Female Involvement, Membership, and Centrality: A Social Network Analysis of the Hartlib Circle", *Literature Compass* 14 (2017) e12388

26. F. L. Maxwell, "Calling for Collaboration: Women and Public Service in Dorothy Moore's Transnational Protestant Correspondence", *Literature Compass*, 14 (2017) e12386

27. Carol Pal, *Republic of Women: Rethinking the Republic of Letters in the Seventeenth Century*, Cambridge University Press, Cambridge, 2012, p. 122

28. L. Hunter, "Sisters of the Royal Society: The circle of Katherine Jones, Lady Ranelagh", *in Women, Science and Medicine 1500-1700*, Lynette Hunter, Sarah Hutton (eds.), Sutton, Stroud, 1997, p. 181

29. Fara, *Op. cit.* (4), pp. 131–132

30. M. DiMeo, *"Such a sister became such a brother": Lady Ranelagh's influence on Robert Boyle, Intellectual History Review*, 25 (2014) 21–36

31. DiMeo, *Op. cit.* (23), pp. 132–133

32. T. D. Whittet, "Some Oxford apothecaries", *Journal of the Royal Society of Medicine* 72 (1979) 940–945

33. Steven Shapin, *Never Pure—Historical studies of science as if it was produced by people with bodies, situated in time, space, culture, and society, and struggling for credibility and authority"*, The Johns Hopkins University Press, Baltimore, 2010, p. 65

34. E. Bourke, ""I would not have taken her for his sister": financial hardship and women's reputations in the Hartlib circle (1641-1661)", *The Seventeenth Century*, 37 (2022) 47–64

35. DiMeo, *Op. cit.* (23), p. 277

36. *Robert Boyle,* Some Considerations Touching the Usefulnesse of Experimental Naturall Philosophy, Oxford, 1664, pp. 81, 113 e 140

37. *Ibid.*, p. 154

38. *Ibid.*, p. 155

39. William R. Newman, Lawrence M. Principe, *Alchemy Tried in the Fire: Starkey, Boyle, and the Fate of Helmontian Chymistry*, The University of Chicago Press, Chicago, 2002, p. 270

40. J. R. Partington, *A Short History of Chemistry*, Dover Publications, New York, 1989, pp. 70–71

41. Robert Boyle, *Occasional Reflections Upon Several Subjects*, edição de 1848; https://babel.hathitrust.org

42. DiMeo, *Op. cit.* (23), p. 11

43. Hunter, *Op. cit.* (28), p. 182

44. M. DiMeo, "Lady Ranelagh's Book of Kytchin-Physic?: Reattributing authorship for Wellcome Library MS 1340", *Huntig Library Quarterly*, 77 (2014) 331–345

45. DiMeo, *Op. cit.* (23), pp. 136, 228, 280

46. M. DiMeo, "Communicating Medical Recipes: Robert Boyle's Genre and Rethorical Strategies for Print", *in* Howard Marchitello, Lyn Tribble (eds.), *The Palgrave Handbook of Early Modern Literature and Science*, Palgrave Macmillan, London, 2017, pp. 209, 216

47. *Ibid.,* p. 215
48. Robert E. W. *Maddison, The life of the Honourable Robert Boyle,* Taylor & Francis, London, 1969, p. 258
49. Hobbes (1588-1679) applied mechanism to political systems. In his *Leviathan* (1651), written during the English Civil War, he proposed a mechanical model of society as a solution to social disorder.
50. Anna Battigelli, *Margaret Cavendish and the Exiles of the Mind*, University Press of Kentucky, Lexington, 1998, p. 39
51. A. Brown, "Lucretius and the Epicureans in the Social and Political Context of Renaissance Florence", *I Tatti Studies in the Italian Renaissance*, 9 (2001) 11–62
52. R. Abraham, S. Roy, "The Atomistic Revival", *World Futures*, 68 (2012) 30–39
53. M. Beretta, "Did Lucretius' Atomism Play any Role in Early Modern Chemistry?", *in* J. Ramón-Sánchez, D. T. Burns, B. Van Tiggelen (eds.), *Proceedings of the 6th International Conference on the History of Chemistry*, Memosciences, Louvain-la-Neuve, 2008, pp. 237–248
54. C. Luthy, "The Fourfold Democritus on the Stage of Early Modern Science", *Isis*, 91 (2000) 443–479
55. S. C. E. Ross, E. Scott-Bauman, Women Poets of the English Civil War, Manchester University Press, Manchester, 2018, p. 212
56. Francis Bacon was an important disseminator of Lucretius' atomism in English scientific circles, which explains its acceptance by the founders of the Royal Society, including John Evelyn.
57. S. Clucas, "Poetic atomism in seventeenth-century England: Henry More, Thomas Traherne and 'scientific imagination'", *Renaissance Studies*, 5 (1991) 327–340
58. D. A. H. Hirsch, "Donne's Atomies and Anatomies: Deconstructed Bodies and the Resurrection of Atomic Theory", *Studies in English Literature, 1500-1900*, 31 (1991) 69–94
59. *new philosophy calls all in doubt, / The element of fire is quite put out, /.......... / And freely men confess, that this world's spent,/When in the planets, and the firmament / They seek so many new; they see that this Is crumbled out again to his atoms./ Tis all in pieces, all coherence gone./... crumbled out again to his atoms. Tis all in pieces, all coherence gone;/All just supply, and all relation:/Prince, subject, father, son, are things forgot.* (verses 205–15)
60. Battigelli, *Op. cit.* (50), pp. 62–63
61. Margaret Cavendish, *Observations upon Experimental Philosophy* (editado por Eillen O'Neil), Cambridge University Press, Cambridge, 2001, p. 85
62. *Ibid.*, p. 129
63. Margaret Lucas Cavendish, *in* Stanford Encyclopedia of Philosophy; https://plato.stanford.edu/entries/margaret-cavendish/
64. S. Clucas, "The atomism of the Cavendish Circle: a reappraisal", *The Seventeenth Century*, 9 (1994) 247–273

65. S. Clucas, J. L. Shaheen, "Part of nature and division in Margaret Cavendish's materialism", *Synthese*, 196 (2019) 3551–3575

66. Margaret Cavendish, *Philosophical and Physical Opinions*, London, 1655, pp. 159–160

67. *Notre Dame Philosophical Reviews*; https://ndpr.nd.edu/news/observations-upon-experimental-philosophy/

68. J. Mason, "The admission of the first women to the Royal Society of London", *Notes Rec. R. Soc. Lond.*, 46 (1992) 279–300

69. S. I. Mintz, "The Duchess of Newcastle's Visit to the Royal Society", *J. Engl. Germanic Philol.*, 51 (1952) 168–176

70. K. Detlefsen, "Reason and Freedom Margaret Cavendish on the Order and Disorder of Nature", *Archiv für Geschichte der Philosophie*, 89 (2007)157–191

71. Cavendish, *Op. cit.* (61), p. 99

72. Quoted by S. Hutton, "In Dialogue with Thomas Hobbes: Margaret Cavendish's natural philosophy", *Women's Writing*, 4 (1997) 421–432

73. *The Diary of Samuel Pepys*, http://www.limpidsoft.com/ipad8/samuelpepys.pdf, p. 2352

74. DiMeo, *Op. cit.* (23), p. 174

75. Quoted by DiMeo (74)

76. Quoted by Margaret Cavendish, *Poems and Fancies with The Animal Parliament*, Brandie R. Siegfried (ed.), Iter Press, Toronto, 2018, pp. 8–9

77. Virginia Woolf, "The Duchess of Newcastle", em *The Common Reader*, Hogarth Press, London, 1925, pp. 74, 78

78. E. Lewis, "The Legacy of Margaret Cavendish", *Perspectives on Science*, 9 (2001) 341–365

79. Carolyn Merchant, *The Death of Nature—Women, Ecology and the Scientific Revolution*, Harper & Row, San Francisco, 1983, p. 270

80. R. Merrens, "A Nature of "Infinite Sense and Reason": Margaret Cavendish's Natural Philosophy and the "Noise" of a Feminized Nature", *Women's Studies* 25 (1996) 421–438

81. Margaret Cavendish, *Sociable letters*, Broadview Editions, Peterborough, 2004, p. 73

82. *Ibid.*, p. 67

83. Quoted by E. Wilkins, "Margaret Cavendish and the Royal Society", *Notes and Records: The Royal Society Journal of the History of Science*, 68 (2014) 245–260

84. S. Hutton, "In Dialogue with Thomas Hobbes: Margaret Cavendish's natural philosophy", *Women's Writing*, 4 (1997) 421–432

85. T. Chao, "Between Nature and Art"—The Alchemical Underpinnings of Margaret Cavendish's Observations upon Experimental Philosophy and The Blazing World, *Euramerica*, 42 (2012), 45–82

86. Cavendish, *Op. cit.* (61), p. 9

87. E. Wilkins, "Margaret Cavendish and the Royal Society", *Notes Rec.* 68 (2014) 245–260

88. K. Dewhurst, "Locke and Sydenham on the teaching of anatomy", *Med. Hist.*, 2 (1958) 1–12

89. D. Boyle, "Margaret Cavendish's Nonfeminist Natural Philosophy", *Configurations*, 12 (2004) 195–227

90. L. T. Sarasohn, "A Science Turned Upside down: Feminism and the Natural Philosophy of Margaret Cavendish", *Huntington Library Quarterly*, 47 (1984) 289–307

91. K. Detlefsen, "Reason and Freedom Margaret Cavendish on the Order and Disorder of Nature", *Archiv für Geschichte der Philosophie*, 89 (2007) 157–191

92. The complete title is "The Description of a New World, Called the Blazing World"

93. S. Åkerman, "Queen Christina's esoteric interests as background to her platonic academies", *Scripta Instituti Donneriani Aboensis*, 20 (2008) 17–37

94. Gabriele Kass-Simon, Patricia Farnes (eds), *Women of Science—Righting the Record*, Indiana University Press, p. 306

95. S. Åkerman, "Queen Christina's Metamorphosis", *in* Judith P. Zinsser (ed.), *Men, Women, and the Birthing of Modern Science*, Northern Illinois University Press, DeKalb, 2005, p. 18

96. Susanna Åkerman, *Queen Christina of Sweden and her Circle: The Transformation of a Seventeenth-Century Philosophical Libertine*, E. J. Brill, Leiden, 1991, pp. 73–74

97. Patricia Fara, *Pandora's Breeches—Women, Science & Power in the Enlightenment*, Pimlico, London, 2004, p. 57

98. Ursula Szulakowska, *The Alchemical Virgin Mary in the Religious and Political Context of the Renaissance*, Cambridge Scholar Publishing, Cambridge, 2017, p. 181

99. James C. Hill, "Johann Glauber's Discovery of Sodium Sulfate—Sal Mirabile Glauberi", *J. Chem. Educ.*, 56 (1979) 593–594

100. Leigh Ann Whaley, *Women's History as Scientists—A Guide to the Debates*, ABC CLIO, Santa Barbara, 2003, pp. 84–89

101. Domna C. Stanton, *The Dynamics of Gender in Early Modern France: Women Writ, Women Writing*, Routledge, Abingdon, 2014, p. 96–97

102. E. Gordon Bodek, "Salonières and Bluestockings: Educated Obsolescence and Germinating Feminism", *Feminist Studies*, 3 (1976) 185–199

103. Whaley, *Op. cit* (100), pp. 82–83

104. M. Tebben, "Speaking of Women: Molière and Conversation at the Court of Louis XIV", Modern *Language Studies*, 29 (1999) 189–207

105. Molière, *Les Femmes Savantes*, Act I, Scene 3

106. L. Tosi, "*La Chymie Charitable et Facile, en Faveur des Dames*, de Marie Meurdrac, une chimiste du XVIIᵉ siècle", C. R. Acad. Sci. Paris, t.2, Série IIc (1999) 531–534

107. Marie Meurdrac, *La Chymie Charitable et Facile, en Faveur des Dames*, Jean d' Hoüry, Paris, 1674

108. Prended Noyce, *Magnificent Minds: Sixteen Pioneering Women in Science and Medicine,* Tumblehome Learning, Boston, pp. 25–26

109. The exact identity of the monarch who was credited with creating the famous water in the late 14th century, which became the most renowned fragrance in Europe until the appearance of Eau de Cologne in the 18th century, remains unknown. Besides being used as a perfume, it was also considered a remedy.

110. S. Feinstein, "La Chymie for Women: Engaging Chemistry's Bodies", *Early Modern Women: An Interdisciplinary Journal* 4 (2009) 223–234

111. Jean-Pierre Poirier, *Histoire de Femmes de Science en France*, Pygmalion, Paris, 2002, pp. 170–176

112. Londa Schiebingen, *The Mind has No Sex? Women in the Origen of Modern Sciences*, Harvard University Press, Cambridge (MA), 1989, p. 113

113. L. O. Bishop, W. S. DeLoach, "Marie Meurdrac—First lady of chemistry?", *J. Chem. Educ.*, 47 (1970) 448–449

114. L. Tosi, "Marie Meurdrac: Paracelsian chemist and feminist", *Ambix* 48 (2001) 69–80

115. Bernadette Bensaude-Vincent, Isabelle Stengers, *Histoire de la chimie*, La Découverte, Paris, 2001, p. 48

116. Poirier, *Op. cit.* (111), pp. 170–179

117. O. Lafont, "Ouvrage de Dame et succès de librairie: les remèdes de Madame Fouquet", *Revue d'histoire de la pharmacie*, 97e année, N. 365 (2010) 57–72

118. Madame Fouquet, *Les Remèdes Charitables de Madame Fouquet*, Jean Certe, Lyon, 1685, pp. 9–13

119. E. L. Keyes, "The Joanna Stephens medicines for the stone A faith that failed", *Bull. N. Y. Acad. Med.*, 18 (1942) 835–840

120. S. Clucas, "Joanna Stephen's medicine and the experimental philosophy", *in* Judith P. Zinsser (ed.), *Men, Women, and the Birthing of Modern Science*, Northern Illinois University Press, Dekalb, 2005, p. 147

121. Marilyn Bailey Ogilvie, Joy Dorothy Harvey (eds.), *The Biographical Dictionary of Women in* Science, vol. 2: L-Z, Routledge, New York, 2000, p. 1227

122. A. J. Viseltear, "Joanna Stephens and the eighteenth century lithontriptics; a misplaced chapter in the history of therapeutics", *Bulletin of the History of Medicine*, 42 (1968) 199–220

123. In present-day terms, that amount would correspond to approximately one million euros.

5

Illuminated and Revolutionaries

It is undeniable that the love of study is far less necessary for the happiness of men than for women. Men have infinite resources for happiness, whereas women lack them completely. Men have many other means of achieving glory. […] however, women, by their condition, are excluded from any kind of glory, and when, by chance, one finds a woman who is born with an elevated soul, there remains for her nothing but study as consolation for all the exclusions and dependencies to which she is condemned by her condition.

ÉMILIE DU CHÂTELET, *Discours Sur le Bonheur* (1779) [1]

During the Enlightenment, chemistry continued to position itself as a practical discipline rather than a theoretical one. Benefiting from centuries-old alchemical techniques and instruments, and simultaneously distancing themselves from the enigmatic speculations of the past, chemists directed their skills toward the preparation of useful products while gradually developing theories about the nature of matter. Near the end of the century, a significant milestone was achieved with the discovery of oxygen and the correct understanding of its role in combustions, calcinations, and respiration. This marked the pivotal transition of chemistry into its modern era, which, at least symbolically, can be said to have occurred in 1789 with the publication in Paris of Antoine Lavoisier's *Elementary Treatise on Chemistry.*

Among the key figures in this chapter, one can find four notable eighteenth-century French women: Émilie du Châtelet, renowned for her mathematical prowess, was intrigued by the nature of fire; Madame d'Arconville, an experimentalist, wrote a treatise on putrefaction; Claudine Picardet, who later became the wife of Guyton de Morveau, Lavoisier's collaborator, made significant scientific translations; and Marie-Anne Lavoisier, the wife of the renowned chemist, served as his secretary and contributed as a scientific translator, illustrator, and editor. Additionally, we encounter Georgiana Cavendish, an English aristocrat and socialite enthusiastic about pneumatic chemistry, and Elizabeth Fulhame, an early British chemist who developed a method for applying metallic films to fabrics and published a study on combustion in 1794. Lastly, we meet the Swedish aristocrat Eva Ekeblad, credited with discovering how to produce ethanol and extract starch from potatoes.

5.1   Ardors and Flames

While not primarily focused on chemistry, Gabrielle Émilie Le Tonnelier de Breteuil, Marquise du Châtelet (1706–1749), deserves a place in this book. A prominent figure in French aristocracy, Émilie du Châtelet (Fig. 5.1), believed in the inherent intellectual capacities of women and their potential to cultivate themselves and achieve autonomy, freedom, and happiness, similar to men [2]. Despite her prefeminist stance, her enduring legacy stems from

Fig. 5.1 *Émilie du Châtelet* (eighteenth century), Maurice Quentin de La Tour. Château de Breteuil

her French translation of Isaac Newton's *Philosophiae Naturalis Principia Mathematica* [*Mathematical Principles of Natural Philosophy*, 1687]. This translation, accompanied by her own annotations and comments, consumed five years of her life, and remains revered as the definitive French reference edition of Newton's seminal work [3].

As a child, Émilie received an exceptional education from her father, Louis Nicolas Le Tonnelier, Baron de Breteuil, even by the standards of a noble girl in her time [2]. At the age of 19, her marriage to Marquis Florent-Claude du Châtelet, a senior army officer, proved to be a disappointment. The couple had little in common: he was dull, devoid of intellectual interests, and primarily focused on his military duties, while she possessed wit and intelligence, with a keen interest in natural philosophy and mathematics, as well as a penchant for social amusements and court festivities [4]. After the birth of her third child in 1733, Émilie entrusted the baby to nannies and resumed her carefree Parisian lifestyle. Like many aristocrats of the era, she had a penchant for frivolity, indulging in expensive clothing, extravagant shoes, and jewelry that cost her husband a fortune. It was during this year that she crossed paths with the famous and controversial François-Marie Arouet (1694–1778), better known by his *nom de plume* Voltaire. As an enlightenment intellectual renowned for his sharp intelligence and criticism of the Church and government, Voltaire had experienced imprisonment and exile from France. By 1733, at the age of 39, he had already spent three years in England, immersing himself in the philosophical ideas of Francis Bacon, Isaac Newton, and John Locke. Voltaire and Émilie became inseparable, engaging in extensive philosophical conversations while also frequenting the opera, theaters, and cabarets of Paris. Voltaire, recognizing her friend's passion for mathematics, introduced her to Pierre-Louis Moreau de Maupertuis, a member of the Académie. Between 1733 and 1734, it was he who provided guidance as she embarked on a largely self-directed learning journey, relying on textbooks. Alexis-Claude Clairaut, also a member of the Académie, later became Émilie's tutor. Under his supervision, she attained an exceptional command of analytic geometry, as well as integral and differential calculus. By the mid-1740s, she stood among the select few in Europe who understood this advanced mathematics [4, 5].

Voltaire's sojourn in England inspired him to write *Philosophical Letters*, a collection of essays about England's institutions, religion, and philosophers. In 1734, when this work, wherein the author openly praised the virtues of the English Parliament, was published in France, it was banned due to the critical views on the French monarchy. Once again facing persecution from the authorities, Voltaire was compelled to hastily leave Paris. In this time of need, Émilie came to his aid by offering him refuge at the Palace of Cirey, her husband's estate located 275 km east of the capital. A few months later, after the palace went through renovations (financed by Voltaire), Émilie joined her companion in ideas and in bed. They welcomed friends and guests who stayed for extended periods. Far from the salons and the voracity of society, Cirey became, between 1735 and 1739, the "place of philosophy and reason," as she called it [4].

It was during this period, specifically in 1738, that the Académie des Sciences announced a prize for the best essay on the nature of fire. While the prevailing belief at the time held that fire was one of the four Aristotelian elements, this topic sparked significant interest and debate. Voltaire took it upon himself to submit his contribution to the Académie, with the assistance of Émilie in conducting various experiments. He concluded that fire was a tangible substance and thus possessed weight. However, driven by her determination to demonstrate otherwise to the Académie, Émilie wrote her own essay titled *Dissertation Sur la Nature et la Propagation du Feu* [*Dissertation on the Nature and Propagation of Fire*, 1744]. It opens as follows:

> Fire reveals itself to us through phenomena so distinct that it is nearly as challenging to define it based on its effects as it is to fully comprehend its nature: it constantly eludes the grasp of our minds. Despite its presence within us, it exists in all the surrounding bodies [6]

Both essays from Cirey failed to win (the Swiss mathematician Leonhard Euler emerged as the victor), but both were published. Émilie du Châtelet's essay marked the Académie's first publication of a woman's work, although she had submitted it anonymously. Today, we can perceive that she was not far from discovering infrared radiation, which would only be known in 1800, thanks to the work of the German-born astronomer and composer, naturalized Englishman, William Herschel (1738–1822).

The Marquise du Châtelet's allusion in her essay to the fire "within us" cannot pass without being appreciated metaphorically, for it is undeniable that she was a woman of diverse and intense fires, whether those of the lust for knowledge or of passion. Her life was extraordinary, both in the way she excelled in mathematics and embraced the issue of women's access to education, and in her roles as wife, mother, friend, and lover. She died at only 42, a week after giving birth to her fourth child, the daughter of the poet Jean-François de Saint-Lambert, her lover. By her deathbed, she had the three men of her life: her husband, Saint-Lambert, and Voltaire. Shortly afterward, the latter, writing to Frederick II of Prussia, would refer to her as "a great man whose only fault was to have been a woman." The persona of Émilie du

Châtelet inspired the Finnish composer Kaija Saariaho to compose the opera *Émilie*, which premiered in 2010 [7].

5.2   Decompositions

Marie-Geneviève-Charlotte Darlus Thiroux d'Arconville (1720–1805), a versatile writer, and translator, as well as an anatomist and experimentalist in the field of chemistry, emerged as a self-taught woman who shone brightly in the intellectual circles of eighteenth-century France (Fig. 5.2). She consistently published her works anonymously, embodying the age-old adage that "a perfect woman is the one you never heard of" [8].

Well aware of the prevailing customs for women venturing into the realm of publishing, she ironically observed that "if their works are bad, they face criticism, and if they are good, they are ignored, leaving them with nothing but the ridicule of having declared themselves as authors" [9]. Approximately half of her written output comprised translations, predominantly from English sources. She fearlessly included her own commentaries and supplements. One of her most renowned works is *Leçons de Chymie* [*Lessons in*

Fig. 5.2   *Madame d'Arconville* (*c.* 1750), Alexander Roslin. Château de Cheverny

Chemistry, 1759], a translation of Peter Shaw's work, supplemented by a 94-page "Introductory Discourse" in which she brilliantly delved into the history of chemistry, exploring topics such as the creation of fire, baking, glass-making, and porcelain production [10]. She also translated a treatise on bone anatomy by Scottish author Alexander Monro, having herself borne the costs of printing the illustrations, whose drawings were done under her supervision [11]. However, it was her treatise *Essai pour Servir à l'Histoire de la Putréfaction* [*Essay in the Service of the History of Putrefaction*, 1766] that brought her the most recognition. She conducted numerous experiments in her private laboratory for this particular work [12].

Marie-Geneviève Charlotte Darlus was born in Paris in 1720, the daughter of the wealthy *fermier général* André-Guillaume Darlus du Tailly. Before her fifteenth birthday, she married Louis-Lazare Thiroux d'Arconville, a Parisian parliamentarian much older than her, and together they had three children. At the age of 23, she contracted smallpox, which left her disfigured and prompted a significant change in her lifestyle. She completely withdrew from her social life and the theater, despite her deep love for it (it is said that she watched Voltaire's tragedy *Mérope* fourteen times in a row) [13]. From that point onward, she dressed and groomed herself as an elderly woman and dedicated herself primarily to study. Her main outings from home were to attend public courses held at the Jardin du Roi, where she learned botany, anatomy, and chemistry. The chemistry course was led by the distinguished Pierre-Joseph Macquer (1718–1784), a friend she greatly admired and through whom she had the opportunity to meet renowned figures of the Parisian scientific community of the time, including the young Antoine Laurent Lavoisier [14]. She also attended chemistry classes given by Guillaume-François Rouelle (1703–1770) in his laboratory [15].

Madame d'Arconville established a chemical laboratory in her residence located in Crosne, a suburban area near Paris. Over a period of more than five years, she dedicated herself to conducting and repeating hundreds of experiments, meticulously recorded in her renowned work, *The Essay on Putrefaction*, which spans 578 pages. This comprehensive work includes ten tables summarizing the conducted tests. For each experiment, predominantly employing raw beef, Madame d'Arconville meticulously documented the date, location, prevailing weather conditions, and ambient temperature. Furthermore, she provided detailed protocols followed during the experiments and, of course, the results, which she compared with those of other authors and offered her interpretations. Among the nearly thirty classes of substances and materials investigated for their antiseptic properties, Madame d'Arconville concluded that cinchona (*Cinchona officinalis*), much like the findings of the renowned

Scottish military doctor John Pringle (1707–1782) [16], exhibited the highest effectiveness. Following cinchona on her list were gallnut powder, powder of palo-santo (*Bursera graveolens*), camphor, peru balsam (*Myroxylon balsamum*), and various salts [17].

Macquer had a significant influence on her, even advising her on the works to be undertaken. The treatise was dedicated to him, albeit anonymously [18]. However, the anonymity was an open secret, as many were well aware of the authorship of the *Essai sur la Putréfaction* [19].

The primary objective of the experimental work leading to Madame d'Arconville's treatise was the search for antiseptic agents to minimize the deterioration processes in animal tissues and, if possible, restore those tissues that were already putrid. Inspired by the suggestions of Macquer and the experiments conducted by John Pringle, the author of *Observations on the Diseases of the Army in Camp and Garrison* (1755), Madame d'Arconville had a strong humanitarian motivation behind her endeavors. She aimed to contribute to the treatment of gangrene in French soldiers wounded during the Seven Years' War (1756–1763). In her *Essay*, she acknowledged her debt to Pringle but also questioned some of his results, attributing any discrepancies to the limited time he had due to his numerous responsibilities [20].

However, in the preface, she reveals a scientific ambition that extended beyond the realm of human benefit. She sought to develop a theory on a subject that intrigued physiologists of the time and that even attracted the attention of encyclopedist Diderot: the transformations of living matter [21]. Her preface states: "We may, therefore, consider putrefaction as Nature's desire... [Nature] only seems to destroy in order to create anew... through her vigilant care nothing is extinguished, all genera intermingle, and pass successively from one kingdom to another by invariable laws which she herself has established, and which she never violates" [22]. Approximately two decades later, the concept of the circulation of matter through the three kingdoms of Nature would be definitively formulated by Lavoisier in the law of conservation of mass.

Like John Pringle, Madame d'Arconville was unaware of the existence and role of microorganisms in putrefaction, a discovery that Theodor Schwann (1810–1882) and Louis Pasteur (1822–1895) would begin to unravel in the following century [23]. The field of biochemistry would later reveal that putrefaction is the result of protein degradation by proteolytic enzymes, leading to the production of biogenic amines with a highly unpleasant odor. However, when stating in the conclusion of her work that she attributed "the preserving power of substances to the obstacle they posed to contact with the external air" [24], she was not entirely far from the truth; and if, in light of what is known today, she had omitted the word "air," she would have gone even further, as saprophytic bacteria are mostly anaerobic.

During the printing process of her treatise, Madame d'Arconville learned that another work on the same subject had just been published by Irish physician David MacBride [25]. Although there was no immediate French translation available (which would only be released in 1771), her honesty and eagerness for the truth prevailed, and promptly, she wrote a supplement in which she acknowledged the new theory as superior. This theory proposed that decomposing materials release *fixed air* (carbon dioxide), which was believed to be the cause of putrefaction. However, it would later be discovered that this theory was entirely incorrect. It is worth noting that *fixed air*, named so because it can be fixed by calcium oxide to form calcium carbonate, was starting to become known in France at that time. Scottish physician and chemist Joseph Black (1728–1799) had discovered it in 1757 and demonstrated that it was the same gas produced during coal combustion, respiration, and fermentation. The introduction of this novelty to France, via Macquer, was made by a Portuguese, João Jacinto de Magalhães, who in 1774 became a member of the Royal Society [17].

Chemist Antoine-François de Fourcroy (1755–1809), who also took an interest in putrefaction, praised Madame d'Arconville's work, considering it on par with contributions from renowned figures such as John Pringle, David MacBride, and Antoine Baumé. Pringle was widely regarded as one of the greatest authorities in the field, MacBride was a pioneer in scurvy treatment, and Baumé is remembered for developing a scale of liquid densities that bears his name. However, after her research on putrefaction, Marie-Geneviève d'Arconville shifted her focus away from experimental work and dedicated herself to charitable activities, as well as translating novels and historical biographies. In 1789, she was widowed, and the onset of the French Revolution brought financial ruin, imprisonment, and the loss of her eldest son. She was eventually released at the end of the Reign of Terror and inherited part of her sister's fortune. She passed away at the age of 85 in 1805 at her home in the Marais.

While Madame d'Arconville focused on the study of putrefaction, her Swedish contemporary Eva Ekeblad (1724–1786) dedicated herself to researching methods for obtaining alcohol and starch from potatoes. It is noteworthy that these two women simultaneously pursued research on two phenomena – putrefaction and fermentation – that were believed to share the same nature and have a purely chemical basis (in the nineteenth century, the renowned German chemist Justus von Liebig (1803–1873) still held this belief).

Fig. 5.3 *Eva Ekeblad* (between 1739 and 1766), Olof Arenius. Finnish National Gallery, Helsinki

Eva Ekeblad, née De la Gardie (Fig. 5.3), upon marrying Count Clas Ekeblad at the age of sixteen, not only received two properties as a wedding gift from her father, the general and statesman Magnus Julius De la Gardie, but also took on the responsibility of managing her husband's estate while he was frequently absent as a member of the Kingdom Council. The countess developed a keen interest in agriculture, which exposed her to the difficult lives of peasants, often plagued by famine.

Originating from the Andes, the potato plant (*Solanum tuberosum*) had been introduced to Sweden in 1658 as an exotic species, showcased in the newly established botanical garden of Uppsala University under the patronage of Queen Christina (see Chap. 4). In 1727, Jonas Alströmer (1685–1761), a pioneer of Swedish agronomy and industry, published a booklet advocating for the nutritional value of the "earth's pear," which was the colloquial name given to the potato. Eva Ekeblad then dedicated herself to developing methods for obtaining both starch and *akvavit* (a popular Nordic distillate containing approximately 40% ethanol) from the tuber. By producing *akvavit* from potatoes, she saved substantial quantities of wheat and barley that could instead be used for bread production. Her efforts significantly contributed to alleviating malnutrition in the country, leading to her official recognition in 1748 with her election to the Royal Swedish Academy of Sciences. In 1751, she further developed a process for bleaching cloth and cotton yarn, and the following year, she demonstrated how potato starch could replace the traditional and toxic arsenic trioxide in cosmetics [26, 27]. Eva Ekeblad's contributions to understanding the nutritional value of the potato preceded the renowned agronomist, nutritionist, and hygienist Antoine Augustin Parmentier (1737–1813), celebrated for introducing this tuber into the French diet. His name became associated with numerous culinary specialties based on the "apples of the earth" (*pommes de terre*).

5.3   New Airs

Let us stay in Parmentier's nation, where two remarkable women, Claudine Picardet (1735–1820) and Marie-Anne Pierrette Paulze (1758–1836), better known as Madame Lavoisier, stood out for their exemplary dedication to chemistry during the Late Enlightenment. Their paths intertwined, and it is believed that the former served as a role model for the latter.

Claudine (Poulet) Picardet, the daughter of a notary in Dijon, quickly developed an interest in scientific subjects, eagerly consuming all sorts of works and magazines of science outreach. At the age of twenty, she married Claude Picardet, an older magistrate who was a member of the Academy of Sciences and Arts and Letters in Dijon. This connection provided her with access to scientific circles and the local upper bourgeoisie. It was in this context that she encountered Louis-Bernard Guyton de Morveau (1737–1816), a young lawyer who was two years her junior. In 1764, he had written a dissertation on public education, proposing a new pedagogical program for French schools after the expulsion of the Jesuits two years prior. This work earned him admission to the Academy of Dijon. However, it was his publication *Digressions Académiques* [*Academic Digressions*, 1772] that garnered him respect within the scientific community. Starting from 1776, he was responsible for the chemistry course at the Academy, attracting students from various nationalities to Dijon each year. Simultaneously, he organized and led a translation group, under whose guidance Claudine Picardet translated works by foreign chemists into French [28]. Initially, a proponent of the phlogiston theory, Guyton de Morveau later adopted the modern chemistry approach championed by Lavoisier [29].

Their shared passion for science and their linguistic abilities must have brought Claudine and Guyton de Morveau together from the very beginning. These language skills proved invaluable when they collaborated on the development of a system of chemical nomenclature, which he introduced in 1782 and which laid the foundation for modern nomenclature. This new system involved naming substances based on their chemical composition rather than relying on subjective criteria such as appearance (e.g., using terms like "antimony butter" for antimony(III) chloride or "vitriol oil" for sulfuric acid) or using names of people or places, as was the convention at the time (e.g., "Glauber's salt" for sodium sulfate or "Epsom salt" for magnesium sulfate) [30]. Claudine, on her part, excelled as a translator of chemical and mineralogical texts. At the suggestion of Guyton de Morveau, she learned German and Swedish in order to translate the works of Carl Scheele (1742–1786) and Torbern Bergman (1735–1784). She also actively participated in his laboratory work, although without formal recognition [31–33].

After having the chance to meet Claudine Picardet in Dijon in 1789, the renowned English agronomist and writer Arthur Young (1741–1820) described her in his book *Travels in France* (1792) as "a very pleasant woman, unaffected, who translated Scheele from German and some of Mr. Kirwan's work from English; a treasure for *Monsieur* de Morveau, as she is not only knowledgeable but also willing to engage in conversations with him on chemical subjects and other instructive or enjoyable topics" [34]. In 1783, the Irish chemist Richard Kirwan had mentioned in a letter to Guyton de Morveau that the latter was "very fortunate to have a lady willing to translate Mr. Scheele's articles," adding that he considered her peerless [35].

The historian Jean-Pierre Poirier remarked that Claudine Picardet belonged to the remarkable group of self-taught women who acquired knowledge out of love for a man [36]. Today, it is difficult to ascertain the extent of her involvement as merely a secretary and translator or if she actively contributed to Guyton de Morveau's scientific work and the development of his ideas, particularly in the realm of new chemical nomenclature and the reform of the system of weights and measures. In any case, she became fully associated with Lavoisier's team of collaborators. After being widowed in 1796, she moved to Paris and resided with her friend, mentor, and lover (now a deputy to the Council of the Five Hundred, professor of inorganic chemistry, and director of the Ecole Polytechnique) whom she eventually married in 1798. Claudine Picardet died in 1820, nearly five years after her husband [37].

The sole surviving portrait of Claudine Picardet, created by an anonymous artist, is believed to have drawn inspiration from Jacques-Louis David's remarkable portrait of the Lavoisier couple (Fig. 5.4). Claudine is depicted

Fig. 5.4 *Marie-Anne Pierrette Paulze Lavoisier and Antoine-Laurent Lavoisier* (1788), Jacques-Louis David. Metropolitan Museum of Art, New York

holding a book, most likely symbolizing her role as a translator. The other individuals portrayed in the painting, are Marie-Anne Paulze Lavoisier, Berthollet, Fourcroy, Lavoisier, and Guyton de Morveau [38].

Regarding Marie-Anne Pierrette Paulze Lavoisier, little can be added to what has already been extensively discussed. Reading her letter to Guyton de Morveau, dated November 16, 1788, is sufficient to understand her remarkable qualities. In this correspondence, she recounts an incident involving an accidental explosion at a gunpowder factory a few days earlier, which nearly claimed the lives of the Lavoisier couple [39]. She writes:

> However, I would prefer not to dwell on such unfortunate matters, as they ultimately lead to the destruction of individuals. Instead, I would like to share with you that I have received a letter from M. de Saussure, who has recently abandoned the phlogiston theory. He credits his shift in belief to the influence of your notes on Kirwan, which has brought him the happiness of embracing the correct doctrine. Nevertheless, it appears to me that in England, proponents of the phlogiston theory are still standing strong. Mr. [James] Keir intends to write an essay on phlogiston. I believe that both of us are exempt from such tasks—I from the duty of translating it, and you from the responsibility of refuting it [40]

After a paragraph in which she expresses optimism about the political developments unfolding in response to the severe financial crisis gripping the nation (which eventually prompted Louis XVI to call for the convening of the Estates General the following year, an event that had not taken place since 1614), Marie-Anne goes on to say:

> M. Lavoisier requests me to convey his cordial regards to you. It would have been his honor to express his gratitude personally, but he did not wish to deprive me of that pleasure. Besides, he is very busy at the moment with a small elementary treatise on chemistry, which will encompass all the procedures for the new experiments, with engraved plates illustrating various apparatuses. As you have rightly suspected, your works will serve as the fundamental basis of his own, and the opportunity to reflect upon them, as well as to quote them, brings him great satisfaction.
>
> Allow me to address a few words to Mme. Picardet, whom I am sure you see every day. I cannot forget the delightful day I had in her company [40]

Furthermore, she makes reference to the recent session of the Académie des Sciences, where the Marquis de Condorcet delivered a brilliant eulogy in memory of the naturalist Comte de Buffon, who passed away during that year. Her closing remarks are:

> I realize now that I have gone too far into the conversation. It is robbing you of precious time, which you use so well to instruct us, and it is making you pay dearly for not living with us. Allow me to end without ceremony, expressing to you all my affection [40]

This correspondence from Madame Lavoisier not only provides insights into her personality and activities but also sheds light on the political climate of France, just eight and a half months before it was to be engulfed in turmoil. Her mention of the Geneva naturalist and geologist Horace-Bénédict de Saussure (1740–799), not only highlights her role as a translator of scientific works, specifically Richard Kirwan's essay on the phlogiston theory (see below), but also emphasizes her role as a disseminator of the emerging field of chemistry through her extensive correspondence with influential figures. Additionally, since the translation of Kirwan's work also contains objections raised by several French chemists, including Lavoisier and Guyton de Morveau, it is appropriate for Marie-Anne Lavoisier to mention Saussure's letter to the latter. Her kind acknowledgment of her husband's treatise (which would be published in the following year, coinciding with the revolution) is admirable, but even more noteworthy is her modesty in omitting her own contribution as the illustrator of this work.

In this context, it is relevant to mention Albertine Necker de Saussure (1766–1841). She was the daughter of Horace-Bénédict de Saussure, sister of the chemist Nicolas Théodore de Saussure (1767–1845), and cousin of the renowned novelist and essayist Madame de Staël (1766–1817). Encouraged by her father, Albertine developed a keen interest in science and experimentation from a young age. She even experienced severe facial burns during an oxygen preparation experiment. Her marriage to the botanist Jacques Necker, who happened to be the nephew of another Jacques Necker (the last Minister of Finance under Louis XVI), appears to have somewhat hindered her laboratory activities. Nonetheless, according to a letter from Guyton de Morveau, he successfully reignited Albertine's passion for chemistry during a visit to the couple. Albertine's correspondence with her father reveals that she conducted experiments in the laboratories of prominent French chemists of that time, including Lavoisier, Fourcroy, and Berthollet [41].

Marie-Anne Lavoisier was well aware of the prestige and renown that she and her husband had garnered. Not only were they esteemed members of the scientific elite, but they also belonged to the upper echelons of French society. It was in the year 1788 that she engaged Jacques-Louis David, her teacher in drawing and painting, to create the famous double portrait measuring 260 by 195 cm (a commission that would be valued at over 200,000 Euros today!). This majestic artwork stood as a symbol of the couple's remarkable accomplishments (Fig. 5.4) [42]. They were truly a brilliant pair, and Marie-Anne herself was hailed at the time with the following verses:

Wife and cousin at the same time,
Sure to love and please,

For Lavoisier, submissive to your laws,
You fulfill both roles,
Of muse and secretary [43]

(The verses were composed by Jean-François Ducis and were originally dedicated to a different lady, Madame de Boufflers [44].)

In 2019, the Metropolitan Museum of Art in New York, where the couple's portrait is exhibited, launched an examination of the artwork. The findings, announced in 2021, proved to be truly remarkable. Through the implementation of analytical techniques such as infrared reflectography and X-ray fluorescence, it was revealed that beneath David's presented composition lies a completely different one [45, 46]. The outcomes of this recent research demonstrate that Marie-Anne was originally depicted wearing a hat adorned with feathers and flowers. Moreover, on the table, there were three opened rolls of paper, and behind it, there was a bookcase with books or files. All these elements have vanished, but in their place, others have emerged in the final version: scientific glass apparatus at Lavoisier's feet and on the table (now concealed under a tablecloth), and on the left side, a portfolio containing Marie-Anne's illustrations for her husband's treatise.

The painting, commissioned from David just a year prior to the revolution, was originally intended to be prominently displayed within the couple's residence. However, upon viewing the initial version of the portrait, which depicted Antoine Lavoisier as a privileged tax collector and Marie-Anne as an aristocrat of the *Ancien Régime*, they must have realized the potential disadvantages it could bring. As a result, the entire composition underwent revisions to transform the portrayal of the couple from a wealthy aristocratic pair to an elegant, hard-working duo engaged in scientific research. In order to safeguard their image, the rolls of paper and the bookcase with tax files, as well as the Marie Antoinette-style hat, disappeared. Additionally, the Louis XVI table was covered with a tablecloth. Stylistically, the transformation can be described as shifting from a rococo scene associated with the old regime to a neoclassical environment symbolizing progress and the future. To convey this message, scientific instruments and equipment were integrated, including oxygen-producing devices on the table, a precision measurement balloon at Lavoisier's feet, and Marie-Anne's drawing boards. Regrettably, the change in message conveyed through the portrait in the face of impending upheaval would not prevent a dramatic outcome.

How did Marie-Anne Paulze, a woman of the late eighteenth century, manage to achieve such notable recognition in her time and secure a place in the annals of scientific history? While it is true that she was the wife of Antoine Lavoisier, that alone would not have been sufficient. Born in 1758 in

Montbrison, in the Loire region, she was the daughter of Jacques-Alexis Paulze, a lawyer at the bailiff's court, and Claudine Thoynet, niece of the influential Abbot Joseph Marie Terray, who served as Louis XV's controller general of finances. Marie-Anne's father, Jacques, owing to his connection with Abbot Terray, gained entry into the *Ferme générale* [47] and secured a position on the board of the Tobacco Commission, which brought him wealth and influence. The untimely death of his wife in 1769 cast a shadow over the family, prompting him to make a decision that would reshape their lives. At the age of not yet twelve, Marie-Anne, who had been residing in a convent (as was customary at the time), returned home to assume the responsibility of managing and bringing solace to the family. She was a charming young girl, petite in stature, with captivating blue eyes, brown hair, and a delicately upturned nose. Alongside her inherent talent and enthusiasm, she possessed a substantial dowry (equivalent to more than 300,000 Euros today). Endowed with such qualities, she soon found herself pursued by a suitor Count d'Amerval. The brother of Baroness La Garde, one of the mistresses of the intimidating Abbot Terray, he was not only already fifty years old but also a dissolute and destitute man. Confronted with this unexpected and delicate situation, Jacques Paulze found himself compelled to engineer an engagement for his daughter that would rescue her from such an unfortunate fate. Carefully observing his fellow *fermiers*, he discerned a possible solution in a twenty-seven-year-old, attractive albeit shy man. This law graduate, with a keen interest in science, had already made significant strides in his career. Apart from his considerable fortune, which facilitated his entrance into the *Ferme générale*, he was a chemist and a member of the Académie des Sciences. His name was Antoine-Laurent de Lavoisier. Marie-Anne was not only drawn to him but also appreciated their conversations, especially those related to scientific subjects. In early 1770, she attended one of his presentations at the Académie, where he discussed the nature of water. Encouraged by Jacques Paulze, Lavoisier finally proposed to the young Marie-Anne, and their wedding was scheduled for December 1771. To everyone's relief, Abbot Terray not only accepted the union but also attended the ceremony. Three years later, in 1774, which coincided with Louis XVI's coronation, Marie-Anne experienced a challenging pregnancy that did not progress favorably. It was also in that same year that Lavoisier published his inaugural work, *Opuscules Physiques et Chi*miques [*Physical and Chemical Opuscula*], in complete dissent from the prevailing chemical doctrine. He defended that air contained an *elastic fluid* (gas) that fixed on metals during their calcination, resulting in the production of *calces* (metal oxides).

Marie-Anne excelled in the art of hospitality, and numerous accounts exist of the couple's lunches and dinners, attended by scientists, philosophers, economists, *fermiers*, and members of high society. Arthur Young portrayed her as follows:

> Madame Lavoisier, a lively, sensitive, scientifically knowledgeable lady, had prepared a *déjeuner anglais* [English-style lunch] with tea and coffee. However, it was her conversation about Mr. Kirwan's *Essay on Phlogiston*, which she is translating from English, and her discussions on various subjects that a cultured woman, who works alongside her husband in the laboratory, knows how to adorn, that was the best part of the meal" [48]

Marie-Anne Paulze's contributions to science extended beyond her role as a hostess. She actively collaborated with her husband, Antoine Lavoisier, by assisting him in tasks such as managing scientific correspondence, particularly with foreign contacts. In her quest to understand Lavoisier's work, she embarked on the study of chemistry. Her initial exploration involved immersing herself in a 1756 edition of Pierre Joseph Macquer's *Éléments de Chymie Théorique* [*Elements of Theoretical Chemistry*, 1749], a work that still adhered to the traditional notion of the four Aristotelian elements and explained combustion through the lens of the phlogiston theory [49]. However, for minds like Antoine Lavoisier's, this worldview was undeniably outdated, rendering Georg Ernst Stahl's (1659–1734) phlogiston theory completely irrelevant. Stahl's theory proposed that combustible substances burned due to the presence of phlogiston – a flammable, invisible, and elusive substance that produced fire when released. According to this understanding, combustible materials like coal, oils, sulfur, and phosphorus were believed to contain an abundance of phlogiston. Conversely, Stahl argued that the *calces* obtained by calcining metals were devoid of phlogiston, as the metals could be regenerated by heating them in the presence of coal (rich in phlogiston), meaning that *metal = calx + phlogiston*. According to the theory, one would expect a decrease in mass during metal calcination as phlogiston escaped. However, experimental evidence revealed the opposite: the mass of the resulting *calx* exceeded that of the original metal. To reconcile this discrepancy, proponents of the Stahlian theory suggested that in such cases, phlogiston possessed negative weight.

Lavoisier acknowledged that, even though incorrect, the phlogistic doctrine had the merit of recognizing that the combustion of substances like sulfur, phosphorus, or coal entailed the same chemical process as the calcination of metals. However, the process involved capturing an *air*. What kind of *air* was it? In 1774, the great French chemist still did not have the answer.

However, in that same year, in October, the Lavoisiers received a visit from the Englishman Joseph Priestley (1733–1804), a staunch follower of Stahl's theory. During dinner, in his hesitant French, he described to them how, from the thermal decomposition of mercuric oxide, he had recently obtained a gas that supported combustions, which he would later designate as *dephlogisti-cated air* [50]. This was the same gas that Carl Scheele had discovered a couple of years prior in Uppsala, which, for the same reasons, he referred to as *fire air*. The following year, Lavoisier successfully identified the component of air responsible for combustion: oxygen, a name he coined because he believed (erroneously) that it was present in all acids (etymologically, "oxygen" means "acid producer").

Lavoisier's appointment to the board of the Administration of Powders and Saltpeter led to the couple moving from rue Neuve-des-Bons-Enfants in April 1776 to the premises of the Petit Arsenal, a stone's throw from the gunpowder depot in the Bastille (a fortress that at the time served as a State prison and whose capture would become a symbol of the French Revolution). Despite its name, the Petit Arsenal was anything but cramped, and much less modest, consisting of a large private apartment, a library, and a laboratory [51]. The couple resided there for sixteen years, during which Lavoisier, utilizing both his own scientific apparatus and existing ones, enhanced by acquisitions he ordered through the State, established an extraordinary laboratory that ranked among the most remarkable of the eighteenth century. His pioneering approach to the need for rigor in scientific measurements led him to equip himself with unprecedented precision instrumentation—barometers, ther-mometers, calorimeters, and scales [52]. His ambitious aim was to reform the entire field of chemistry, and Marie-Anne's proficiency in English, which he lacked, proved invaluable. Not only did she correspond with notable figures like Joseph Priestley and Henry Cavendish (1731–1810), but she also trans-lated key works of chemical literature herself. Two of these works were Richard Kirwan's *An Essay on Phlogiston, and the Constitution of Acids* (1787) and *Of the Strength of Acids and the Proportion of Ingredients in Neutral Salts* (1790). These translations were part of Lavoisier and his group's strategy to dismantle the phlogiston theory, which enjoyed strong support from the respected Irish chemist.

The initial French edition of the *Essay on Phlogiston* (1788) does not specify the translator's name (this oversight was later rectified). However, it credits Lavoisier and his four closest collaborators (Morveau, Laplace, Monge, Berthollet, and Fourcroy) as the authors of the appended notes challenging the phlogiston theory. Some believe that Claudine Picardet may have contrib-uted to the translation [53]. Additionally, the translation includes a preface and three minor corrections to Kirwan's work, which were also made by Marie-Anne. These circumstances warrant the recognition that Lavoisier's wife played a more significant role than a mere translator in actively attacking the phlogiston doctrine [54]. Her correspondence with individuals like Saussure further attests to this fact (as seen above).

Her artistic talents granted her the authorship of the illustrations featured in her husband's *Traité Élémentaire de Chimie* [*Elementary Treatise of Chemistry*, 1789] (Fig. 5.5), in addition to two drawings depicting the laboratory experiments he conducted while investigating human respiration and perspiration (Fig. 5.6) [55].

The *Elementary Treatise* includes the law of conservation of mass, also known as Lavoisier's law, asserting that in a chemical reaction, the total mass of the reacting substances equals the total mass of the products formed. Additionally, it introduces a new, logical, and systematic chemical nomenclature designed for universal understanding. This achievement, building upon the initial groundwork laid by Guyton de Morveau, as mentioned earlier, reflects the enlightenment's tendency toward classification and systematization (it was, after all, the century of the Swedish Carl Linnaeus, the father of modern taxonomy, and the *Encyclopédie* by the Frenchmen Diderot and d'Alembert). Lavoisier's treatise, however, gives prominence to a new element:

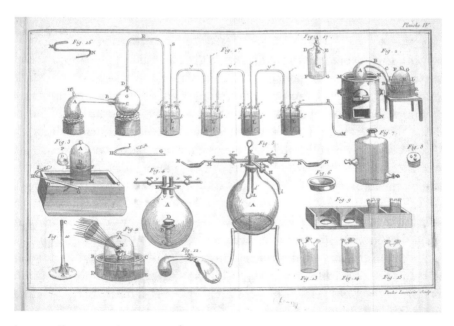

Fig. 5.5 Illustration from *Traité Élémentaire De Chimie*, Cuchet, Paris, 1789. Courtesy of the Science History Institute

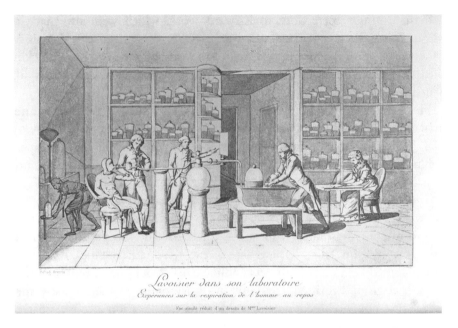

Fig. 5.6   Lavoisier in the laboratory investigating human respiration (*c*.1790). Drawing by Marie-Anne Paulze (self-portrait at right, seated at a table taking notes). Courtesy of the Edgar Fahs Smith Memorial Collection, Kislak Center for Special Collections, Rare Books and Manuscripts, University of Pennsylvania

oxygen. This pivotal discovery dismantled the prevailing phlogiston theory, despite its lingering proponents. To commemorate this milestone, Marie-Anne orchestrated at the Arsenal a ceremony, symbolically burning the works of Stahl while dressed as a priestess—a kind of phlogiston auto-da-fé [56].

The dissemination of the new chemistry also relied on personalities like Antoine Fourcroy, a collaborator of Lavoisier, who, apart from his dedication to science, became involved in politics as well. During his tenure as minister, he took the initiative to establish the National Museum of Natural History in 1793. The institution's roots can be traced back to the King's Garden, originally founded as the Royal Garden of Medicinal Plants by Louis XIII in 1635. Within its grounds, a new amphitheater became the venue for Fourcroy's public chemistry courses, drawing enthusiastic audiences, including a significant number of female attendees. In his work *Voyage au Jardin des Plantes* [*Journey to the Plant Garden*, 1797], Louis-François Jauffret, a French educator, writer, and advocate of science popularization, underscores the value of Nature walks as a means to acquire knowledge. In the book, Jauffret narrates the journeys of two friends as they explore these Parisian gardens and attend one of Fourcroy's lessons:

As the session had already begun, we found all levels of the amphitheater occupied; we had only room in the topmost row, and even there we were not seated. Fourcroy spoke with his usual ease, and although we were at quite a distance from him, we could see him perfectly well. Filled with admiration, Gustave observed Fourcroy before his gaze encompassed the entire audience. Surprised to see a circle of ladies at the end of the amphitheater, he whispered to me: "What are those ladies doing here? Do they want to learn chemistry?—Yes, my friend, they are the descendants of *Philaminte* and *Bélise* [...]. In stark contrast to the countless frivolous women whose conversations revolve solely around fashion and clothing, as well as the ordinary women preoccupied with domestic matters, these ones, seemingly at ease, engage exclusively in discussions centered on scientific subjects. They have just learned from Fourcroy that *nitre*, or saltpetre, is now called *nitrate of potash*, and that cooking salt is called *muriate of soda*; they come to learn how to decompose into the elements; to convert water into air, and air into water; to know the two components of ordinary air, which are oxygen and nitrogen gas.—Do they understand all this?—Almost all of it. Fourcroy possesses the ability to express himself very clearly, making science accessible to all. Let us listen to him for a few minutes; I bet you will understand him" [57]

(*Philaminte* and *Bélise* are two female characters depicted in Molière's comedy *Les Femmes Savantes*; see Chap. 4.)

The developments of the revolution were not favorable to Lavoisier. His vast fortune and the connection to the *Ferme générale*, along with numerous enemies, most notably the physician and politician Jean-Paul Marat (1743–1793), ultimately led to a tragic end. In August 1793, he was compelled to depart from the Arsenal, and subsequently, on November 28, he was arrested. From his prison cell on December 19, he wrote to his wife, whose desperate endeavors to secure his freedom were proving to be in vain:

You inflict upon yourself much pain, my good friend; so much weariness of body and mind that I cannot share. Do not let your health deteriorate, that would be the greatest of misfortunes. I have lived a long life, filled with happiness since the day I met you. You have contributed and contribute to it every day with your proofs of dedication. Finally, I leave behind me memories of esteem and consideration. My mission is complete, but you still have the right to count on a long career. Do not waste it. Yesterday, I sensed you were sad. Why should you be sad? Know that I have accepted everything with resignation and consider any losses as gains. Furthermore, the hope of our reunion remains ever-present and, in the meantime, your visits bring me precious moments of joy [58]

After the execution of her husband and father on May 8, 1794, along with 36 other *fermiers*, Marie-Anne herself was arrested on June 14 and endured 65 days of imprisonment. She faced a period of severe financial hardship, mitigated by the generosity of a former servant. Determined to reclaim her belongings, including the couple's portrait, she persevered and eventually found success, gradually regaining her position in society. From 1800 onward, her salon, one of the last refuges of the culture of the Enlightenment that survived the Terror, began to shine once again [59]. It was there that emerging young scientists (Fourier, Gay-Lussac, etc.) had the opportunity to make contact with Lavoisier's former collaborators (Berthollet, Lagrange, Laplace, etc.). And no small number of foreign scientists stopped by (Humboldt, Davy, Berzelius, etc.), paying homage to Madame Lavoisier [60].

In 1805, fully aware of her husband's historic contribution to the development of chemistry and determined to ensure his legacy, Marie-Anne took the initiative to publish his unfinished work *Mémoires de Chimie*. This private edition, exclusively intended for personal distribution and not available for sale, included an introduction penned by herself, which complemented the existing fragments. When she died in 1836, there were still 1100 copies of the *Mémoires* remaining. These, along with other volumes from her library (including *The Lusiads*, an English translation of Camões' epic poem [61]), were eventually sold at auction [62].

Following her husband's death, Marie-Anne received marriage proposals from Pierre Dupont (1739–1817), an economist and politician, who was a longtime friend of the couple and an enduring admirer of hers (they might have been lovers). However, her second marriage was to Benjamin Thompson, Count Rumford (1753–1814), an American-born physicist, inventor, and adventurer known for his contributions to thermodynamics. He had earned the title of count in Munich for his outstanding services to Bavaria in various fields such as military affairs, science, agriculture, and public health. They formalized the union in October 1805, but their marriage was far from a blissful one. Despite four years of courtship and shared travels in Switzerland and Bavaria, their relationship swiftly deteriorated after their wedding. Various factors may have contributed to the discord between them. One potential source of contention was Marie-Anne's stipulation, included in the marriage contract, that she be recognized as Madame Lavoisier de Rumford [63]. On the other hand, Rumford, with his opinionated and presumptuous nature, had envisioned marriage as a means to achieve a luxurious lifestyle, desiring a submissive wife by his side. In contrast, Marie-Anne cherished her autonomy and appreciated being the mistress of her own life. One of the many conflicts between the couple was recounted by Rumford himself in a letter to his daughter Sally in 1807, who resided in the United States of America. Marie-Anne had organized a reception at home for a large group of guests without informing her husband. The latter, taking the incident as a provocation, locked the iron gates to the property, keeping the key to himself. When the guests arrived and gathered at the entrance to the garden, all the hostess could do was to talk to them over the high wall that separated them. She then went to water her husband's revered flowers with boiling water. Interestingly, Sally seemed unimpressed by her father's accounts of the incident. When visiting him in Paris in 1811 (already after the couple's amicable separation), she expressed admiration for her former stepmother's personality [62].

There is an intriguing and almost cinematic account of the final years of Marie-Anne Lavoisier's life, written by M.A. Delahanty, the grandson of a former friend of the Lavoisiers who also worked for the *Ferme générale*:

> I remember her with a feeling of respectful gratitude, mixed, however, with a certain sense of terror. It was under her auspices that my brother Gustave, and I, two young men, made our entrance into the world of society, which was not exactly a pleasure, for almost every Sunday we had to go, from the Place de la Sorbonne, to visit Madame de Rumford. We arrived at Rue d'Anjou, in front of a horseshoe-shaped wall cut by the gate of the vast garden, in the middle of which stood the house. The doorman opened a small side door at the bend of the horseshoe, and we entered the large wing leading to the building, while a loud bell announced our arrival. There, our hearts began to pound, but it was too late to retreat. By the time we reached the lobby we had bravely recovered. An old, balding lackey dressed in an old French-style livery welcomed us with a warm smile and led us through the winter garden to the door of the salon, accompanying us with an intimidating solemnity.
>
> The first thing that caught one's eye upon entering that room was, on a panel to the right, a large canvas of David depicting M. and Mme. Lavoisier. M. Lavoisier, sitting at a table on which you could see chemistry instruments, was dressed in Louis XVI style; behind her husband and leaning on the back of his armchair, the young, powdered Mme. Lavoisier was dressed entirely in white.
>
> We walked toward the fireplace, while admiring the magnificent portraits, and arrived in front of a sofa at the end of which curled up a sort of old Turk. This old Turk was all that remained of the beautiful young woman David had

painted: she was Madame de Rumford, with a figure masculinized by aging and combed and dressed in a bizarre way.

She received us with her benevolent abruptness, made us sit down and started a conversation about our studies and our pleasures, topics that perhaps did not hold much interest for her. After a few minutes of conversation, it often happened that, rising suddenly from the sofa, she would go and stand with her back to the fireplace, like men. She would lift her skirts behind her to the height of her garters and quietly warm her huge calves. Soon after, she would politely dismiss us, which we didn't need to hear twice.

She used to give beautiful dances, which amused us much more than the visits we made to her, despite the active surveillance and strictness with which she sent us from the buffet to the ball. She also gave great dinners, but because of our age we were rarely invited. One day, however, we received an invitation, almost an order. It was a dinner for a small number of guests, served in the winter garden. The restricted group consisted of M. François Arago, M. de Humboldt, M. Cuvier and the young Count Napoléon Daru, a promising graduate of the École Polytechnique who was beginning to be noticed for his work in the Chamber of Peers. If by placing us in the presence of such personalities she intended to impress our young imaginations, then she did not miss her goal. Such a dinner would never escape our memory [64].

Marie-Anne's generosity extended beyond mere philanthropy; she also showed gratitude to those who had supported her during the most challenging times, including some of her former employees. She was equally attentive to her nieces, especially her favorite, the countess de Gramont, to whom she gave a handsome wedding dowry. However, there exists a specific episode that sheds further light on her character. Being well aware of her niece's fondness for cakes, Madame Lavoisier always made sure they were abundantly present on the table whenever she invited her for tea. On one occasion, Madame Lavoisier went a step further and instructed one of the maids to wrap up the leftover cake after the tea and place it in the countess' carriage, as a surprise. After bidding their farewells, as the countess stepped into the carriage and discovered a greasy package resting on the satin seat, she promptly shook it out without hesitating. The result was that her aunt, who had observed everything from the window, would never leave her a single penny in inheritance [62]. This apparent harshness of Madame Lavoisier can, however, be viewed with some leniency. The profound impact of her husband and father's simultaneous execution, the abandonment by numerous friends, the experience of imprisonment, the enduring hardships, and the arduous quest to reclaim her possessions had left indelible scars on her. While she ultimately regained wealth and respect, the days of loving and pleasing forever belonged to the past. Her passing in 1836 at the age of 78, occurring during a period when France had reinstated monarchy, marked the end of an era.

As early as 1817, while visiting Paris, Mary Somerville (see next chapter) expressed her frustration at the challenge of finding French women who engaged in conversations beyond the topic of fashion:

> [...] I could not avoid being struck by the difference between the accomplishments of French and English Ladies. Among all I have met only one pretended to know a little music and it was poor indeed, two drew a little, in language and science I met with none except Mme Biot.... Certainly I found none of the high cultivation of mind and elegance of manners so constantly seen in England, not among the higher classes alone but widely diffused through the nation. Dress is a great object among the French ladies and forms a frequent subject of conversation [65].

5.4   Inhalations and Reductions

Let us confine our attention to the United Kingdom and explore the involvement of two female figures during the late eighteenth century in the field of chemistry: Georgiana Spencer (1757–1806) and Elizabeth Fulhame (about whom little information is available). Georgiana, an English aristocrat descended from the Spencer family, which also included Diana Spencer, Princess of Wales (separated by seven generations), exhibited a strong passion for scientific disciplines, notably chemistry and mineralogy (Fig. 5.7). As a young woman, Georgiana displayed an exceptional dedication to learning and attained an education that far surpassed the norm for aristocrats of her time. Married to William Cavendish, Duke of Devonshire, she embraced the life of socialite and became a prominent figure in the realm of fashion. She was described as "possessing extraordinary charm and endowed with generous impulses and good nature, but, on the other hand, incredibly reckless and frivolous." Prone to gambling, extravagant spending, and engaging in extramarital relationships, she embraced a lifestyle that was not uncommon among the nobility of the eighteenth century. Nevertheless, the details of her tumultuous personal life will not be explored here [66].

Through her marriage to the fifth Duke of Devonshire, Georgiana established familial ties to Henry Cavendish (grandson of the second Duke), the

Fig. 5.7 *Georgiana Cavendish* (1783), Thomas Gainsborough. National Gallery of Art, Washington

natural philosopher who discovered hydrogen in 1766. This union also linked her to Margaret Cavendish, the Duchess of Newcastle (see Chap. 4). Despite the duke's objections, Georgiana initially visited Henry Cavendish's laboratory but was eventually prohibited doing so, reportedly because he was a laborer [66]. Georgiana went on to create a small laboratory at Chatsworth House, the family's country estate, where she dedicated herself to the practice of chemistry. In London, she frequented the Royal Academy to consult scientific treatises and conducted her experiments in a room at Devonshire House, the family's palace in the capital, situated on Piccadilly [67].

In December 1793, while in Bristol, she visited the laboratory that the physician Thomas Beddoes (1760–1808) had recently set up for the investigation of the therapeutic effects of inhaling gases, several of which had been identified and characterized not long ago (carbon dioxide, hydrogen, oxygen, nitrous oxide, etc.). Beddoes was so struck by the Duchess of Devonshire's grasp of chemistry that he felt compelled to share it in a letter to his colleague, Erasmus Darwin (the paternal grandfather of Charles Darwin), saying that "she had revealed a knowledge of modern chemistry far superior to that which he imagined any English duchess or lady could possess." Georgiana's fascination with the potential of pneumatic medicine led her to make a second, extensive visit to Beddoes' laboratory. It was during this occasion that the idea of establishing a hospital institution, the Medical Pneumatic Institution, began to take shape. Eager to garner support for Beddoes' project, she approached Joseph Banks (1743–1820), the English naturalist and botanist who then served as the President of the Royal Society. Despite her efforts, Banks refused to lend his support. The reasons extended beyond scientific concerns and delved into politics, as Beddoes was known to espouse the ideals emerging from the French Revolution. Georgiana's appeal to James Watt (1736–1819), the renowned Scottish inventor celebrated for his advancements in steam engine technology, also failed to convince Banks. Despite the lack of success of these attempts, the Pneumatic Institution was created in 1799. Much of the scientific equipment found within it had been specifically designed by James Watt. His profound admiration for pneumatic medicine stemmed from the tragic loss of his daughter to tuberculosis at a young age. Heading the laboratory was Humphry Davy, a young English chemist (1778–1829), who dedicated himself, among other investigations, to studying the effects of nitrous oxide (laughing gas) on his own body. While the results achieved at the Pneumatic Institution fell short of initial expectations, they laid the groundwork for the future of inhalation anesthesia [66].

Limited information is available regarding Elizabeth Fulhame, already dubbed the "forgotten genius" [68]. What is known is that she was married to a physician named Thomas Fulhame, who received his education at Edinburgh University. In 1794, Elizabeth Fulhame authored an influential work titled *An Essay on Combustion* (Fig. 5.8), which was translated into German four years later and had an American edition in 1810 [69]. In this publication, she details and provides commentary on numerous chemical experiments she conducted, primarily focusing on the application of metals onto fabrics. This objective is explicitly mentioned in the preface:

> The possibility of making cloths of gold, silver, and other metals by chymical processes, occurred to me in the year 1780; the project being mentioned to Doctor Fulhame and some friends, was deemed improbable. However, after

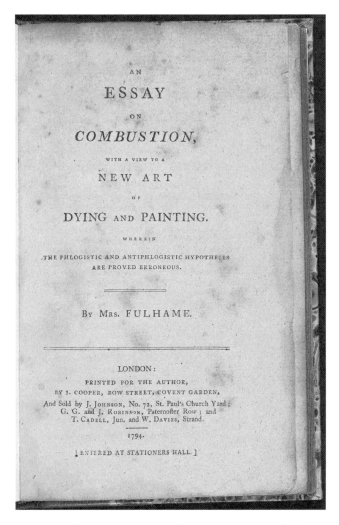

Fig. 5.8 Frontispiece of *An Essay on Combustion*. Courtesy of the Science History Institute

some time, I had the satisfaction of realizing the idea in some degree by experiment [70].

Further on, she states:

Some time after this period, I found the invention was applicable to painting, and would also contribute to facilitate the study of geography; for I have applied it to some maps, the rivers of which I represented in silver, and the cities in gold. The rivers appearing as it were in silver streams, have a most pleasing effect on the sight, and relieve the eye of that painful search for the course and origin of rivers [71].

She reveals that she does not mince words. Her directness is reminiscent of a feminist manifesto rather than a scientific work:

I may appear presuming to some, that I should engage in pursuits of this nature: but averse from indolence, and having much leisure, my mind led me to this mode of amusement, which I found entertaining and will hope be thought inoffensive by the liberal and the learned. But censure is perhaps inevitable; for some are so ignorant, that they grow sullen and silent, and are chilled with horror at the sight of anything that bears the semblance of learning, in whatever shape it may appear; and should the spectre appear in the shape of woman, the pangs which they suffer are truly dismal [72].

Toward the end of such a vibrant statement, she adds:

But happen what may, I hope I shall never experience such desertion of mind, as not to hold the helm with becoming fortitude against the storm raised by ignorance, petulant arrogance, and privileged dullness [73].

In a more serene tone, the ending of the preface hints at the importance of her work for chemistry:

Finding, the experiments could not be explained on any theory hitherto advanced, I was led to form an opinion different from that of M. Lavoisier, and other great names. Persuaded that we are not to be deterred from the investigation of truth by any authority however great, and that every opinion must stand or fall by its own merits, I venture with diffidence to offer mine to the work, willing to relinquish it, as soon as a more rational appears [74].

Elizabeth Fulhame's interest was focused on metal ion reduction reactions that lead to deposition of metals onto fabrics. Throughout several years, she investigated the reduction of various metal ions (gold, silver, mercury, copper, tin), using an array of reducing agents including hydrogen, phosphorus, potassium sulfide, hydrogen sulfide, phosphine, and coal. Her most notable scientific achievement revolved around investigating the feasibility of obtaining metals through chemical reduction in aqueous solutions. Additionally, she delved into the study of light-induced reduction of metal ions [75]. In her experiments, she encountered phenomena that posed challenges to the prevailing theories of combustion and reduction at the time. Her references to the work of chemists such as Antoine Lavoisier, Pierre Macquer, Richard Kirwan, and Carl Scheele, among others, reveal a remarkable knowledge of the science of her time. While she leaned toward Lavoisier's antiphlogiston perspective, she didn't hesitate to critique his views, by stating that "the antiphlogistic account of calcination and reduction is no less complex, erroneous, and repugnant in the face of the simplicity of nature" [76].

She goes on to recall that "according to M. Lavoisier, the oxygen, which oxygenates combustible bodies, is sometimes derived from vital air; sometimes from acids; sometimes from water; and sometimes from metallic oxides, etc." [76]. However, she shows herself strongly convinced that "the hydrogen of water is the only substance that restores oxygenated bodies to their combustible state; and that water is the only source of the oxygen that oxygenates combustible bodies" [77]. Her "numerous, painstaking, and terribly tedious experiments" [78] led her to note that water had the ability to accelerate the combustion of carbon, transforming it into carbon dioxide. This observation foreshadowed the concept of catalysis, a term coined by Berzelius in 1835, derived from the Greek words "*kata*" and "*lyein*," meaning "down" and "to loosen," respectively [78–80].

Elizabeth Fulhame's work garnered attention and recognition within the scientific community at the time. Count Rumford, Benjamin Thompson (see above), replicated her experiments on the photoreduction of ionic gold and acknowledged obtaining similar outcomes. Similarly, Jean-François Coindet (1774-1834), a Geneva physician renowned for his pioneering use of iodine in goiter treatment, wrote a positive 27-page critical review of her work [78]. The criticism came mainly from Priestley, a proponent of phlogiston [81]. The English chemist, who is said to have witnessed Fulhame's experiments in London, considered her theory as "absurd and fictional as the story of the phoenix itself."

The Irish chemist William Higgins (1763–1825), one of the first proponents of atomic theory in the nineteenth century, took a more critical stance. He accused her of plagiarism, or at the very least, of disregarding his prior work. Higgins specifically referred to his own discovery that iron undergoes rusting exclusively in the presence of water [79–83]. Fulhame's book received widespread acclaim in the United States, leading to her election as an honorary member of the Philadelphia Chemical Society [84].

References and Notes

1. Émilie du Châtelet, *Discours Sur le Bonheur*, Rivages, Paris, 2014. pp. 52–53
2. P. Jonas de Lima Piva, F. Tamizari, "Luzes femininas: a felicidade segundo Madame du Châtelet", *Estudos Feministas*, 20 (2012) 853–868
3. J. P. Zinsser, "Translating Newton's 'Principia': The Marquise du Châtelet's Revisions and Additions for a French Audience", *Notes and Records of the Royal Society of London*, 55 (2001) 227–245
4. D. E. Musielak, "The Marquise du Chatelet: A Controversial Woman of Science", 2014; https://arxiv.org/ftp/arxiv/papers/1406/1406.7401.pdf
5. J. P. Zinsser, "Mentors, the Marquise Du Châtelet and Historical Memory", *Notes and Records of the Royal Society of London*, 61 (2007) 89–108
6. Émilie du Châtelet, *Dissertation Sur la Nature et la Propagation du Feu*, Prault, Paris, 1744 pp. 1–2
7. João Paulo André, *Poções e Paixões – Química e Ópera*, Gradiva, Lisboa, 2018, p. 53
8. Patrice Bret, Brigitte Van Tiggelen (eds.), *Madame d'Arconville (1720-1805): Une femme de lettres et de sciences au siècle des Lumières*, Hermann, Paris, 2011, p. 7
9. Quoted in Marie Geneviève Charlotte Thiroux d' Arconville, *Mélanges de Littérature, de Morale et de Physique*, tome 1, Amsterdam, 1775 pp. 370–371
10. Peter Shaw, *Leçons de Chymie, Propres à Perfectionner la Physique, le Commerce et les Arts*, J. T. Herissant, Paris, 1759
11. Londa, Schiebinger, *The Mind Has No Sex? Women in the Originis of Modern Science*, Harvard University Press, Cambridge (MA), 1989, p. 249
12. Tax collector on behalf of the king.
13. Jean-Pierre Poirier, *Histoire des Femmes de Science em France- Du Moyan Age à la Révolution*, Pygmalion, Paris, 2002, p. 265
14. *Ibid.*, p. 266
15. C. de Milt, "Early Chemistry at Le Jardin du Roi", *Journal of Chemical Education*, 18 (1941) 503–509
16. S. Selwyn, "Sir John Pringle: hospital reformer, moral philosopher and pioneer of antiseptics", *Med. Hist.*, 10 (1966) 266–274
17. Poirier, *Op. cit.* (13), p. 268
18. M.-G.-C. Thiroux d'Arconville, *Essai pour servir à l'histoire de la putréfaction*, Didot Le Jeune, Paris, 1766 n.p.
19. Bret and Van Tiggelen, *Op. cit.* (8), p. 94
20. John Pringle is traditionally considered the "father of military medicine".
21. Elisabeth Bardez, "Au fil de ses ouvrages anonymes, Madame Thiroux d'Arconville, femme de lettres et chimiste éclairée", *Revue d'Histoire de la Pharmacie*, LVII (2009) 255–266
22. d'Arconville, *Op. cit.* (18), pp. xi–xii

23. Seymour S. Block (Ed.), *Disinfection, Sterilization, and Preservation*, Lippincott Williams & Wilkins, Philadelphia, 2001, p. 12
24. d'Arconville, *Op. cit.* (18), p. 546
25. D. Macbride, *Experimental Essays*, A. Millar, London, 1764
26. Jeri Freedman, *Women of the Scientific Revolution*, The Rose, New York, 2017, pp. 26–28
27. E. Bodensten, "A societal history of potato knowledge in Sweden c. 1650–1800", *Scandinavian Journal of History* (2020) 1–22
28. R. C. Mocellin, "Louis-Bernard Guyton de Morveau e a revolução química das Luzes", *Scientiæ Studia*, 10 (2012) 733–58
29. E. M. Melhado, "Oxygen, Phlogiston, and Caloric: The case of Guyton", *Hist. Stud. Phys, Sci.* 13 (1983), 311–334
30. Cathy Cobb, Harold Goldwhite, *Creations Of Fire: Chemistry's Lively History From Alchemy to the Atomic Age*, Basic Books, New York, 2001, p. 163
31. Marelene Rayner-Canham, Geoffrey Rayner-Canham, *Women in Chemistry: Their Changing Roles from Alchemical Times to the Mid-Twentieth Century*, Chemical Heritage Foundation, 2001, Philadelphia, p. 23
32. T. E. Thorpe, *Essays in Historical Chemistry*, Macmillan, London, 1902, pp. 62–63
33. Poirier, *Op. cit.* (13), p. 278
34. Quoted in Mathilda Betham-Edwards (ed.), *Arthur Young's Travels in France*, George Bell & Sons, London, 1892 (4th edition), p. 222
35. Quoted in reference 31
36. Poirier, *Op. cit.* (13), p. 279
37. Patrice Bret, "Les promenades littéraires de Madame Picardet La traduction comme pratique sociale de la science au XVIII siècle", *in* P. Duris (ed.), *Traduire la Science – Hier et aujourd' hui*, Maison des Sciences de l'Homme d'Aquitaine, Pessac, 2008, pp. 125–152
38. Marco Beretta, *Imaging a Career in Science: The Iconography of Antoine Laurent Lavoisier*, Science History Publications/USA, Canton, 2001, p. 100.
39. It was an experiment involving a gunpowder composition that, instead of saltpetre, contained potassium chlorate, a substance recently discovered by Claude Berthollet.
40. Quoted in M Lucien Scheler, "Deux lettres inédites de Mme Lavoisier", *Revue d'histoire des sciences*, 38 (198) 121–130
41. Rayner-Canham, *Op. cit.* (31), p. 24
42. Poirier, *Op. cit.* (13), p. 295
43. Madison Smartt Bell, *Lavoisier no Ano Um – O Nascimento de Uma Nova Ciência Numa Época de Revolução*, Gradiva, Lisboa, 2005, p. 24
44. *Épouse et cousine à Ia fois,/Sûre d'aimer et de plaire, /Pour Boufflers, soumis à vos lois, / Vous remplissez Ies deux emplois, /Et de muse et de secrétaire.* Jean-François Ducis, *Œuvres posthumes*, G. Doyen, Paris, 1826, p. 239
45. S. A. Centeno, D. Mahon, F. Carò, D. Pullins, "Discovering the evolution of Jacques-Louis David's portrait of Antoine-Laurent and Marie-Anne Pierrette Paulze Lavoisier", *Heritage Science*, 9 (2021)

46. N. Kenney, "Progressive scientists, or high-flying elitists? The Met unlocks a secret behind a famous Jacques-Louis David portrait", *The Art Newspaper*, 1 September 2021
47. "General Treasury" in English. It refers to a company granted by the State for the collection of indirect taxes.
48. Quoted in Mathilda Betham-Edwards (ed.), *Arthur Young's Travels in France*, George Bell & Sons, London, 1892 (4th edition), p. 94
49. Jean-Pierre Poirier, *La Science et l'Amour – Madame Lavoisier*, Pygmalion, Paris, 2004, p. 25
50. *Ibid.*, pp. 26–27
51. C. Viel, "Lavoisier avait-il un laboratoire autre que celui de l'Arsenal?", *Revue d'Histoire de la Pharmacie*, 307 (1995) 369–373
52. Joe Jackson, *A World on Fire: A Heretic, an Aristocrat, and the Race to Discover Oxygen*, Viking, New York, 2005, p. 182
53. Poirier, *Op. cit.* (49), p. 90
54. K. Kawashima, "Madame Lavoisier et la traduction française de *l'Essay on phlogiston* de Kirwan", *Rev. Hist. Sci.* 53 (2000) 235–263
55. M. Beretta, "Imaging the Experiments on Respiration and Transpiration of Lavoisier and SÉguin: Two Unknown Drawings by Madame Lavoisier", *Nuncius*, 27 (2012) 163–191
56. Arthur Greenberg, *A Chemical History Tour: Picturing Chemistry from Alchemy to Modern Molecular Science*, Wiley – Interscience, New York, 2000, p. 146
57. Louis-François Jauffret, *Voyage au Jardin des Plantes, contenant la description des galeries d'histoire naturelle, des serres où sont renfermés les arbrisseaux étrangers*, Paris, 1797 pp. 36–38
58. *Revue des Deux Mondes* tome 79 (1887) p. 902
59. M. Beretta, "Una musa per la chimica? Marie Anne Paulze-Lavoisier e la scienza del suo tempo", *in* Raffaella Simili (org.), *Scienza a Due Voci*, Leo S. Olschki, Firenze, 2006, p. 102
60. *Ibid.* p. 103
61. *Catalogue de livres faisant partie de la bibliothèque de feu de Madame Lavoisier*, Galliot, Paris, 1836, p. 42
62. D. I. Duveen, "Madame Lavoisier 1758-1836", *Chymia*, 4 (1953) 13–29
63. M. Guizot, *Madame de Rumford (1758-1836)*, Crapelet, Paris, 1841, p. 29
64. M. A. Delahante, *Une famille de finance au XVIII siècle*, Paris, 2 (1881) 546–549
65. Quoted in Elizabeth C. Patterson, *Mary Somerville and the Cultivation of Science, 1815–1840*, Martinus Nijhoff Publishers, The Hague, 1983, p. 22
66. N. A. Bergman, "Georgiana, Duchess of Devonshire, and Princess Diana: a parallel", *Journal of the Royal Society of Medicine*, 91 (1998), 217–219
67. Marelene Rayner-Canham, Geoffrey Rayner-Canham, *Pionnering british women chemists: their lives and contributions*, World Scientific, London, 2020, pp. 5–6
68. K. J. Laidler, "Lessons from the History of Chemistry", *Acc. Chem. Res.* 28 (1995) 187–192

69. The complete title is *An Essay on Combustion, with a view to a new art of dying and painting. Wherein the phlogistic and antiphlogistic hypotheses are proved erroneous*
70. Elizabeth Fulhame, *An Essay on Combustion*, J. Cooper, London, 1794, p. iii
71. *Ibid.* p. iv
72. *Ibid.* p. xi
73. *Ibid.* p. xii
74. *Ibid.* p. xiii
75. Rayner-Canham, *Op. cit.* (31), p. 29
76. Fulhame, *Op. cit.* (70), p. 7
77. Fulhame, *Op. cit.* (70), p. 8
78. D. A. Davenport, K. M. Ireland, "The ingenious, lively and celebrated Mrs. Fulhame and the Dyer's hand", *Bull. Hist. Chem.* 5 (1989), 37–42
79. Greenberg, *Op. cit.* (56), p. 156–161
80. K. J. Laidler, A. Cornish-Bowden, "Elizabeth Fulhame and the discovery of catalysis: 100 years before Buchner", *in* A. Cornish-Bowden (ed.), *New Beer in an Old Bottle. Eduard Buchner and the Growth of Biochemical Knowledge*, Universitat de València, 1997, pp. 123–126
81. K. L. Neeley, M. A. Bashore, "Esteem, regard, and respect for rationality: Joseph Priestley's female connections", *Bull. Hist. Chem.*, 30 (2005) 77–90
82. Quoted in reference 78
83. Rayner-Canham, *Op. cit.* (31), p. 30
84. J. C. Linker, "The Pride of Science: Women and the Politics of Inclusion in 19th Century Pennsylvania", *Pennsylvania Legacies* 15 (2015) 8–11

6

Authors and Readers

The Chemical conviction
That Nought be lost
Enable in Disaster
My fractured Trust -

The Faces of the Atoms
If I shall see
How more the Finished Creatures
Departed me!

EMILY DICKINSON, *The Chemical Conviction* (nineteenth century)

Despite the prevailing barriers to acquiring a formal education, eighteenth-century women had access to a variety of textbooks tailored to disseminate knowledge of natural philosophy. Often perceived as having fragile minds, these women only needed minimal knowledge of the era's progress. Titles such as *Astronomy of the Ladies*, *Newtonianism for Ladies*, and *Chemistry for Women* – the latter written by the Italian author Giuseppe Compagnoni – were among these books. However, the work that had the most profound impact, remaining a topic of discussion to this day, is *Conversations on Chemistry*, authored by the English writer Jane Marcet and published in the early nineteenth century. Shortly after, in the United States of America, Almira Hart Lincoln Phelps stood out as an author of science textbooks for girls. Moving into the twentieth century, the American chemist Mary Fieser excelled in writing university textbooks.

J. P. André, *Sisters of Prometheus*, https://doi.org/10.1007/978-3-031-57136-7_6

6.1   Complementary But Not Equal

Despite the eighteenth century often being referred to as the Age of Enlightenment or the Age of Reason, formal education for women did not significantly advance during this period. Nevertheless, it became fashionable for women in social salons to engage in conversations that encompassed themes of natural philosophy [1]. They were not expected to possess in-depth knowledge but rather to have a basic understanding of important discoveries and theories, for which a few dedicated books were available [2]. Women's magazines also featured articles on philosophical-natural themes, along with lectures and public courses that women could attend [3]. Initially, these books had a highly literary style, often in verse, which gradually diminished in the latter half of the century. The dialogue format, appreciated since Galileo's *Dialogue on the Two Principal Systems of the World* in 1632, remained common and continued into the first decades of the nineteenth century. Over time, women's scientific literature adopted a more "masculine" tone, shedding the once-prevalent poetic embellishments. For example, in *Astronomie des Dames* [*Astronomy of the Ladies*, 1786], the author and astrologer Jêrome de Lalande, while not completely rejecting the model established a century earlier by Bernard Le Bovier de Fontenelle in *Entretiens Sur la Pluralité des Mondes* [*Conversations on the Plurality of the Worlds*, 1686], introduced innovative features. Instead of combining scientific content with stories and dialogue, as his predecessor had done, Lalande chose a simple and direct style to address a female audience. However, at the time, many found this approach too dry and dull [4].

Toward the end of the eighteenth century, despite the increasing acknowledgment of the potential equality between women and men in terms of intellectual capabilities and their contributions to society, as exemplified by figures like the Marquis de Condorcet (1743–1794), discussed in Chap. 1 of *Sisters of Prometheus—From the New Woman to Nobelity in Chemistry*, the prevailing belief in female inferiority persisted. Among the advocates of feminine subalternity were figures such as Jean-Jacques Rousseau (1712–1778), despite the renowned philosopher from Geneva having been one of the early champions for equality among individuals in Western culture. However, he evidently excluded the female sex. In his famous work *Émile* (1762), Rousseau outlined the educational objectives for the protagonist, the young Émile, emphasizing autonomy and preparation for public service. In contrast, he assigned the role of a passionate yet submissive wife and a demanding mother to Sophie, Émile's fiancée:

Hence, it is essential to educate women in relation to men. Their primary duties include satisfying and being useful to men, earning their love and respect, nurturing them from childhood, caring for them in old age, offering advice, consolation, and ensuring a pleasant and happy life. These lifelong obligations define a woman's role and must be ingrained in her from an early age [5].

Rousseau not only believed that women were less intelligent than men, but he also claimed that trying to educate them in the same way as men was a total waste of time. As abstraction was, by nature, inaccessible to women, according to his perspective, all the creative components associated with science fell short of their capabilities. At their best, these capabilities manifested in the domains of the concrete and the practical. Even when it came to popularized scientific subjects, he believed that women's competence was quite limited [6, 7]. Physically, mentally, and socially, they were complementary to men, not their equals. Moreover, this belief was reinforced by the medical knowledge of the era, which highlighted the distinctions between the genders [8]. In the *Encyclopédie*, in 1765, a significant portion of the "skeleton" entry was devoted to comparing men and women. Its author concluded that variances in the skull, vertebral column, clavicles, sternum, coccyx, and pelvis demonstrated that "the purpose of women is to bear and nourish children" [9]. The writer Pierre-Joseph Boudier de Villemert (1716–1801) also held the belief that men and women were not physically and intellectually equivalent but rather complementary opposites. Accordingly, he acknowledged the existence of scientific domains that were particularly well-suited to women's intelligence [10]. However, in his *L'Ami des Femmes* [*The Friend of Women*, 1758], he argued against women engaging in speculative and technical sciences, as he believed that such "thorny tasks" could weaken their spirit. According to him, figures like Émilie du Châtelet (see Chap. 5) or Anne Le Fèvre Dacier (1647–1720), a renowned translator and commentator of classical authors, were exceptions that should be admired rather than imitated. The studies de Villemert recommended to women were those leading to moral improvement, namely, history and even physics. However, he specifically emphasized the practical aspect of physics, as it allowed for the admiration of the wonders of Nature. While abstract studies were considered beyond the reach of women, those that appealed to the imagination, which was believed to be a feminine quality, were deemed perfectly suitable. Activities like painting, poetry, and music, both in terms of practice and appreciation, were also regarded as highly suitable for women [10]. Chemistry was another science that de Villemert considered appropriate for them, given its practical applications in domestic tasks. In relation to this, in the previous chapter, we saw how Madame d'Arconville, with her treatise on putrefaction, and, above all, Claudine Picardet and Marie-Anne Lavoisier, with their translations, made important contributions to the chemical literature of the eighteenth century.

However, it was botany that acquired the connotation of a female science par excellence, given its close association with home medicine and decorative preoccupations. It was also considered to have the advantages of fostering intellectual discipline and to be healthy, as it was mainly performed outdoors. The Swedish botanist Carl Linnaeus (1707–1778), by basing his classification of plants on their respective sexual organs, disrupted this *idyllic* perception. Nonetheless, despite the initial shock, botany continued to be viewed as a science particularly suitable for young women, as it did not weaken their spirit and allowed, like no other, the glorification of divine creation [11, 12]. The female fascination with this science came to be known as "botanomania." Interestingly, Rousseau himself undertook the task of instructing a lady in botany, as demonstrated by his eight *Lettres Élémentaires Sur la Botanique a Madame de L*** [*Elementary Letters on Botany to Madame de L***], written between 1771 and 1774 [13].

6.2   By the Philosopher's Hand

With very few exceptions, one of the significant contributions women made to science in the eighteenth century was its popularization, albeit indirectly. They appeared as fictional characters in numerous science outreach books, thereby contributing to making science more accessible to the public. Often, the titles of these works suggested they were intended exclusively for women. However, in reality, they aimed at a wider range of readers [14]. The influential work that sparked this trend was *Entretiens Sur la Pluralité des Mondes* (1686) by Le Bovier de Fontenelle (mentioned above). This book, presented in the form of dialogues, consists of six lessons on Copernican heliocentrism and Descartes' philosophy, delivered to the Marquise de G*** by a modern philosopher. Through this work, Fontenelle, the perpetual secretary of the Académie des Sciences of Paris and a great communicator of science, unveiled a new facet of the natural philosopher. The latter was no longer seen solely as an academic or scholastic master surrounded by male disciples, but rather a sympathetic guide who led women along the paths of science, making it understandable to those who, like them, generally lacked any knowledge on the subject [15]. As for the choice of a female character in the *Entretiens*, the author clarifies in the preface: "I have introduced in these Conversations a woman who is being taught and is completely unfamiliar with such matters

[…] in order to make the work more acceptable and to inspire ladies through the example of someone who, without any prior knowledge of science, attentively listens to what she is being taught, organizing the vortices and the worlds in her mind without confusion" [16]. It is worth noting that a few years prior to the publication of this work, the treatise *De l'Égalité des Deux Sexes* [*On the Equality of the Two Sexes*, 1673] was also published in France. Its author, François Poullain de La Barre, a Cartesian priest and philosopher, argued that men and women possessed identical intellectual capacities [17].

The *Entretiens* of Fontenelle achieved great success, with 10 French editions and several translations during the author's lifetime, serving as a model for other works. One of the earliest examples was *Il Newtonianismo Per le Dame* [*Newtonianism for Ladies*, 1737] by Francesco Algarotti, a Venetian who dedicated it to the influential French writer. Originally published in Milan, this work went through multiple editions and translations. Algarotti had partly written it in France, where he resided with the duo Émilie du Châtelet and Voltaire at the Cirey Palace in 1734 (as discussed in Chap. 5) [18]. However, Algarotti did not have to leave the Italian Peninsula to encounter other notable women who excelled in physics and mathematics during that time. One such figure was Laura Bassi (1711–1778), who became known as the Minerva of Bologna (Fig. 6.1). Born into a family of lawyers, she underwent 7 years of private education during her adolescence. In 1732, she earned her doctorate from the University of Bologna, becoming the first woman to obtain a doctorate and teach physics at a European university. In 1776, at the age of 65, she was granted the chair of experimental physics, which she held until the end of her days [18, 19]. In *Il Newtonianismo*, the scientific education of the young debutante, the Marquise de E***, begins after she inquires about a poem celebrating a female scientist of the era. The poem, written by Algarotti himself in 1732, extols of the scientific intellect of Laura Bassi.

The deliberate scientific naivety portrayed in the dialogues of this book facilitates the teaching of the various topics to an untrained public, irrespective of their gender, contrary to what the title may suggest. The use of familiar analogies drawn from the Marquise's domestic world—such as the household, the garden, everyday objects, etc.—significantly aids in *her* comprehension of essential scientific principles. In this way, this fictional aristocrat actively contributes to the promotion of scientific education among the literate population [20]. The publication of this work was not without controversy; it was initially listed in the index of prohibited books. However, subsequent editions were revised to address these concerns, ultimately resolving the issue [15].

Fig. 6.1 *Laura Bassi* (eighteenth century), Carlo Vandi. Sistema Museale di Ateneo, Bologna

Let us now briefly mention Laura Bassi's scientific pursuits, which suggest her potential involvement in the field of chemistry, as indicated by some of her correspondence. One of her many scientific correspondents was the young Italian physicist Alessandro Volta (1745–1827), who, upon learning of her interest in electrical phenomena, discussed with her the experiments conducted using a firing device relying on the reaction between *metallic air* (hydrogen) or *marsh air* (methane) with *dephlogisticated air* (oxygen), which could be triggered by an electric spark. Bassi, who received a eudiometer [21] from the Italian physicist, toxicologist, and anatomist Felice Fontana (1730–1805) in 1775, engaged in the study of gases and even wrote a dissertation on the subject. Her correspondence with Fontana and Volta suggests her involvement in the debate between Lavoisier, the proponent of the new chemistry, and Priestley, a defender of the phlogiston theory (as discussed in Chap. 5). In Italy, Fontana supported Priestley, while Lavoisier was supported by the naturalist and physiologist Lazzaro Spallanzani (1729–1799), who was Bassi's cousin. Unfortunately, since Bassi passed away without completing her studies and the dissertation no longer exists, we cannot determine her position in that discussion [22, 23].

Over the course of six decades, a number of science popularization works emerged in the vein of Algarotti's *Il Newtonianismo*. As noted by researcher Fiamma Lussana, there was a dual effect regarding the popularization of science among women during the enlightenment period. On one hand, it became fashionable for them to acquire scientific knowledge that they could showcase during intellectual discussions in the elegant salons of the time. On the other hand, these works also served to alleviate some of the tension stemming from the well-known debate that started in 1723 at the Accademia dei Ricovrati in Padua [24]. This academy, with which Galileo Galilei had been associated since its foundation in 1599, stood out as one of the few academies that included women among its members, albeit in honorary positions. One of them was the Venetian Elena Lucrezia Cornaro Piscopia (1646–1684), professor of mathematics at the University of Padua and the first woman in the world to receive a doctorate in philosophy. Another famous female member of the academy was Madeleine de Scudéry (see Chap. 4). 1723 was the year the newly elected president initiated an (all male) debate on the women's question, namely, whether women should or should not be admitted to the study of sciences and liberal arts. Only 6 years later, women would have the right to reply, and although inconclusive, the debate did contribute to a reflection on women's access to public spheres [25].

The series of scientific dissemination works that followed *Il Newtonianismo* included titles such as *Il Filosofismo delle Belle* [*The Philosophism of the Beautiful Ones*, 1753], by Giovanni Cattaneo; *Il Libro Per le Donne* [*The Book for Women*, 1757], by Enea Gaetano Melani, an abbot who believed in the natural propensity of women to study science; and *La Filosofia Per le Dame* [*Philosophy for Ladies*, 1777], by an anonymous author [24, 26, 27]. Toward the end of the century, continuing the tradition of the dialogue between a philosopher and a lady wishing to acquire scientific knowledge, an important work disseminating the new chemistry was published in Venice: *La Chimica Per le Donne* [*Chemistry for Women*, 1796], by Giuseppe Compagnoni (Fig. 6.2a).

Compagnoni (1754–1833) was a Roman constitutionalist, journalist, and writer. He is credited with creating the *tricolore*, the Italian flag. His work *La Chimica Per le Donne*, presented in two volumes, consists of 59 letters addressed to Marianna Rossi, a countess of Ferrara, followed by five more letters in an appendix. The first volume contains 37 letters and is preceded by an informational note for readers, which opens with the following words:

At the beginning of this century, all educated men talked about *attraction*. In these last few years, everyone is talking about the *new chemistry*.

a

b

Fig. 6.2   (a) Giuseppe Compagnoni and (b) Vincenzo Dandolo

Next, the author states that, despite the existence of multiple works dealing with chemistry, there is still none that presents

> […] the elements of this new science in such a simple and clear manner that can educate curious without the complex language that generally […] evokes repulsion or generates boredom and annoyance [28].

Continuing to justify the purpose of his book, Compagnoni expresses his intention to assist those who seek to gain

> […] a clear understanding of the new science, as nothing can be comprehended or articulated without first acquiring the vocabulary it employs […],

further emphasizing that this knowledge will be

> […] indispensable even for those who […] have extensively read about the experiments and discoveries of the eminent chemists prior to Lavoisier [29].

However, as his basic training was not in science, he sought the help of Vincenzo Dandolo (1758–1819), a chemist and a dear friend (Fig. 6.2b). In 1791, Dandolo had translated Lavoisier's *Traité Élémentaire de Chimie* into Italian (see Chap. 5), and 4 years later, shortly after the tragic death of the French chemist, Dandolo published his own book titled *Fondamenti della*

Scienza Chimico-Fisica [*Foundations of the Science of Chemistry-Physics*], wherein he introduced the new chemical terminology. Compagnoni expressed his gratitude to his friend in these words:

> If I can have any merit for the idea of providing my fellow citizens with a course in this science suited for people of all classes, undoubtedly, the one who supplied the materials deserves even more. Nor here, oh virtuous Dandolo, shall my friendship or your modesty deprive you of the just praise you deserve. Not only have you provided me with the subject for this work through your remarkable book [*Fondamenti*], which brings honor to Italy, but you have also patiently and courteously supported and guided me at every step […] [30].

Interestingly, in this introduction, there is no specific mention of the female gender. On the contrary, the author's intention is to reach a broad audience encompassing both men and women, as well as individuals from all social backgrounds.

In the opening lines of the first epistle ("Opportunity of this work" and "Importance of modern chemistry"), Compagnoni (the philosopher) addresses his female counterpart with some incredulity:

> So you expect me, Countess, to discuss chemistry with you! But why would you require it? Women excel as teachers in the field of chemistry [31].

After praising the proficiency of women in handling distillation, Compagnoni acknowledges the seriousness of his addressee's request and recalls her reasons for wanting to learn chemistry. She had mentioned that, since chemistry had become a trendy science, she believed she had the right to learn it. Additionally, she expressed her frustration with the constant criticism women faced regarding their attires, hairdos, games, and fashionable recreations, which perpetuated men's intolerance. The lady had also suggested that by making studying fashionable, it would no longer be perceived as a grave sin. In response to these points, Compagnoni says to her:

> Countess, I shall not contest your argument. It would be impolite, indiscreet, and perhaps even unfair; the thirst for knowledge is a shared affliction among both sexes of our species. How could it be an exception in your case? You are absolutely right, and I have no objection to your request [32].

After discussing the nature of chemistry and alluding to alchemy and its esoteric character, he explains that

[…] men, after having long deluded themselves chasing shapeless phantasms, created by a delirious imagination, have finally discovered their true interest, and are now more devoted than ever to the practical sciences, of which chemistry ranks among the foremost [33].

And he continues:

I have here, my illustrious friend, the very object to which your requests call me. This will be the focal point of our correspondence […]. The undertaking I shall embark upon is none other than *Chemistry for Women*. Half a century ago, another Italian wrote *Il Newtonianismo* for your gender. Such audacity inspires me. Given the novelty and significance of the subject matter, I may potentially garner favor with those who rightly appreciate Algarotti's elegance [34].

The second letter to the countess contains a "brief history of chemistry," starting with Moses and the golden calf and concluding with Lavoisier. The third letter introduces several fundamental definitions. From this point forward, the correspondence expands beyond the immediate realm of chemistry (gases, acids, new nomenclature, etc.), encompassing subjects such as the impact of light on plants, respiration, and even topics less aligned with Lavoisier's focus, such as meteorology, aurora borealis, tides, and earthquakes.

The Countess of *La Chimica Per le Donne* is the last in a series of fictional female characters that appeared throughout the eighteenth century as vehicles for the dissemination of scientific themes. As historian Paula Findlen observes, the inclusion of women in such works did not stem from their being the intended target audience, but rather from their effective representation of the general public that typically did not consume science books [35]. However, the "garden *filosofessa*" was not without critics, who accused her of contributing to the degradation of science by domesticating it [14]. This perspective is evident in the satirical work *Disgrazie di Donna Urania ovvero Degli studi femminili* [*The Misfortunes of Donna Urania*, 1793] by Robbio di San Raffaele. According to this book, women's intellectual pursuits were deemed incompatible with their roles as wives and mothers. The protagonist of these misadventures is depicted as a pedantic, exhibiting traits reminiscent of a *précieuse* (as described in Chap. 4):

[…] her favorite pursuit was the study of the natural sciences, which engrossed her to such an extent that her house […] quickly became filled with an array of machinery, instruments, and contraptions, all in pursuit of unraveling Nature's most mysterious secrets.

Urania embarked on various collections, but her eagerness to enrich them all at once resulted in their shortcomings. Each collection lacked essential components and instead overflowed with trivial and commonplace items of little value or rarity. The dovecote, once a training ground for her grandfather's city pigeons, now proudly bore the title of an observatory, serving as a means for the Lady to capture astronomical phenomena while observing the stars. In a corner of the garden, a rudimentary botanical garden took shape. A handful of devices, albeit flawed and adorned with gilded bronze leaf, awaited the addition of another forty to truly earn the designation of a Cabinet of Physics. Halls, corridors, and even bedrooms, stripped of their usual decorations, succumbed to the scientific fervor that claimed them for scholarly pursuits, filling them with philosophical apparatus.

Even the kitchen faced the invasion, although the cook, displaying rare presence of mind, voiced objections, asserting that the realm of flavors held a theoretical and practical school of its own, and that he himself was a skilled practitioner in this domain. However, the pantry could not escape the transformation; it begrudgingly yielded to the establishment of a chemical laboratory, where the Lady, refusing to be a mere observer, frequently busied herself with the duties of Cyclopesa [36].

6.3   Jane's Conversations

The period following Lavoisier's death witnessed significant advances in quantitative analysis, notably with the contribution of chemists such as the German Martin Heinrich Klaproth (1743–1817), the French Louis-Nicolas Vauquelin (1763–1829), and the Englishman William Hyde Wollaston (1766–1828), which led to the discovery of new metallic elements. In 1799, French chemist Joseph-Louis Proust (1754–1826) established the law of definite proportions, according to which when two or more elements combine to form a compound, their masses are always in a constant ratio. In turn, the Englishman John Dalton (1766–1844) played a crucial role by publishing the first table of atomic masses in 1803. His work laid the foundation for the modern atomic theory, which was far from immediately accepted. Besides the usual resistance to novel ideas, it faced challenges due to errors in atomic masses, uncertainty surrounding compound formulas, and a fundamental unawareness of the difference between atoms and molecules. The law of combining volumes of gases, stated in 1808 by the Frenchman Louis Joseph Gay-Lussac (1778–1850)—that the volumes of gases involved in a chemical reaction can be expressed as simple whole number ratios—could have been instrumental, but instead, it only augmented confusion. The Italian Amedeo Avogadro

(1776–1856), who foresaw the difference between atoms and molecules, proposed that equal volumes of gases at the same temperature and pressure would contain equal numbers of particles. Nevertheless, his hypothesis went unnoticed for almost five decades. The breakthrough came at the Karlsruhe Conference in 1860 when Stanislao Cannizzaro, a chemist from Sicily, distributed his pamphlet titled *Sunto di un Corso di Filosofia Chimica* [*Outline of a Course in Chemical Philosophy*], shedding light on Avogadro's pioneering work.

Meanwhile, in Stockholm, Jöns Jacob Berzelius (1779–1848) made notable advancements in isolating new elements, starting with cerium discovered in 1803, followed by selenium, thorium, silicon, zirconium, and titanium, and in determining atomic masses. He also deserves credit for the development of modern chemical notation, as he suggested the use of the current chemical symbols in 1811 [37, 38]. In Bristol, at the Pneumatic Institution, where the effects of various gases on human beings were investigated between 1799 and 1802 (as described in Chap. 5), Humphry Davy (1778–1829) emerged as a prominent figure. In the midst of the era of discovery of electricity (marked by Alessandro Volta's presentation of the first electric battery in 1800), Davy made remarkable advancements in the field of electrolysis, isolating several elements between 1807 and 1808, including potassium, sodium, barium, strontium, calcium, and magnesium. Furthermore, he successfully isolated boron in 1808 and chlorine in 1810 [39]. Davy's public lectures at the Royal Institution in London, which commenced in 1801, featured captivating experimental demonstrations that enthralled the audience. These lectures played a pivotal role in inspiring a woman, Jane Marcet, to write one of the most influential chemistry textbooks ever published: *Conversations on Chemistry* (1806).

Born in London in 1769, Jane Haldimand Marcet (Fig. 6.3) was the daughter of an English mother and a Swiss father, who was a highly successful banker and businessman. Growing up in a stimulating environment, Jane and her 11 siblings were provided with a rich education, guided by some of the best tutors of the era. In line with their father's beliefs, both boys and girls were given equal opportunities for learning, resulting in Jane and her siblings receiving instruction in subjects such as history, Latin, natural philosophy, as well as art, music, and dance [40].

Jane, who from an early age revealed a great appetite for knowledge, possessed a shrewd and pragmatic spirit that enabled her to have an almost literal understanding of words. An anecdote is often recounted to illustrate her way of thinking: when she was a child, she heard her grandmother complaining about a watch that didn't work properly, which she would gladly exchange for

Fig. 6.3 Jane Haldimand Marcet. Courtesy of Edgar Fahs Smith Collection, Kislak Center for Special Collections, Rare Books and Manuscripts, University of Pennsylvania

a guinea. Jane, unabashedly, the next day presented her with a proud guinea and a half, just as much as she had bargained for the watch at a watchmaker's. To her great astonishment, instead of the praise she thought she deserved, she heard a strong reprimand from the old lady, who thus taught her that not everything that is said should be interpreted literally. Throughout her life, Jane would maintain that one shouldn't say one thing when they mean another [41].

From an early age, she had the opportunity to meet important political, intellectual, and social figures who frequented her parents' home. At the age of 15, she experienced the sudden loss of her mother, who succumbed to a complicated childbirth at the age of 39. Within the family, a decision was made that she would take on the responsibility of organizing the Haldimand social evenings. It was during one of these events that she met Alexander John Gaspard Marcet (1770–1822), a Swiss man whom she would marry in 1799. Together, they had four children. Alexander Marcet, originally from Geneva, had sought refuge in England in 1794 during the tumultuous period of the French Revolution. In 1797, he earned his medical degree from the University of Edinburgh, and 2 years later, he was admitted to the Royal College of Physicians. That was also the year when he married Jane. In 1805, he played a role in the establishment of the Medical and Chirurgical Society of London, an institution that fostered connections with physicians and chemists from various countries, including Jacob Berzelius. During this period, Alexander focused his efforts on analyzing mineral waters, gaining valuable experience for his work in the chemical analysis of animal fluids and urinary calculi. His significant contributions led to the publication of *An Essay on the Chemical History and Medical Treatment of Calculous Disorders* in 1819, which brought him recognition and acclaim [40, 42]. His marriage to Jane led to the Haldimand-Marcet house becoming a meeting place for prominent figures in the scientific world, including chemists such as Berzelius, Wollaston, and Davy [40]. In addition, Jane hosted her own salon, aimed at fostering connections among intellectual women. Her circle of acquaintances included the polymath Mary Somerville (1780–1872), a renowned science writer who deserves recognition in this chapter, as well as the sociologist, social activist, and writer Harriet Martineau (1802–1876), and the novelist Maria Edgeworth (1768–1849). Jane Marcet and the latter are reported to have formed a friendship after Maria expressed her gratitude to the author of *Conversations* for allegedly saving her sister's life. She had accidentally ingested an acid, and in the midst of the emergency, someone recalled reading in Jane's book about the use of magnesia, a suspension of magnesium hydroxide that serves as an antacid, for neutralizing acids [43].

Jane and Alexander Marcet were frequent attendees of Davy's public lectures at the Royal Institution. Enthralled by the presentations, Jane not only aspired to understand the lectures in their entirety but also endeavored to replicate the practical demonstrations at home, with the evident support of her husband. These endeavors sparked her realization of the importance of making knowledge about the principles of chemistry and its societal impact accessible to a broader audience, particularly women. Motivated by this insight, she decided to write a textbook, culminating in the publication of *Conversations on Chemistry* in 1806. Initially published anonymously, the book nonetheless indicated to be written by a woman, with the name of its author only being revealed in the thirteenth edition in 1837 [44]. The profound pedagogical purpose of *Conversations on Chemistry* is immediately evident from the outset of its preface:

In venturing to offer to the public, and more particularly to the female sex, an Introduction to Chemistry, the author, herself a woman, conceives that some explanation may be required; and she feels it the more necessary to apologise for the present undertaking, as her knowledge of the subject is but recent, and as she can have no real claims to the title of chemist [45].

One cannot help but notice the contrast between the diplomacy exhibited in this justification and the impatience expressed in the preamble of Elisabeth Fulhame's essay on combustion, which was published 10 years earlier (see Chap. 4).

Conversations on Chemistry, originally published in two volumes—the first dealing with what was then referred to as "simple bodies" and the second with "compound bodies"—is structured as a series of 26 informal dialogues. These dialogues take place between a governess, Mrs. B., and her two pupils: the reserved 13-year-old Emily and the spontaneous 10-year-old Caroline. The use of the dialogue format had been revived in England through popular science works such as *Botanical Dialogues Between Hortensia and Her Four Children* (1797) by Maria Elizabetha Jacson, and *Domestic Recreation, or Dialogues Illustrative of Natural and Scientific Subjects* (1805) by Priscilla Wakefield [46]. Jeremiah Joyce also contributed to this trend with his *Scientific Dialogues* (1800–1803), which covered mechanics, astronomy, hydrostatics, pneumatics, optics, magnetism, electricity, and galvanism [47]. It is worth mentioning that a year after Jane Marcet's work was published, Joyce released *Dialogues in Chemistry* (1807), featuring an anonymous tutor and two boys named Charles and James [48].

In the preface to *Conversations*, the author justifies her choice of the dialogue format as she considers it "a most useful auxiliary source of information; and more especially to the female sex, whose education is seldom calculated to prepare their minds for abstract ideas, or scientific language" [49]. Throughout the work, the main concepts and principles of chemistry are conveyed in a direct and simple manner, along with the significant advancements made in the field. This includes the discoveries made by prominent figures such as Galvani, Volta, Franklin, Rumford, Priestley, Oersted, Berzelius, Berthollet, Cavendish, Lavoisier, Davy, and many others. Jane Marcet's exposure to a great deal of late eighteenth- and early nineteenth-century chemistry had given her an overview of the state of development of this science, something that, with all due differences, few modern-day educators could match with respect to nowadays chemistry [50]. While the first edition of *Conversations on Chemistry* predates the publication of John Dalton's *New System of Chemical Philosophy* (1808), it is noteworthy that scientists like Humphry Davy did not

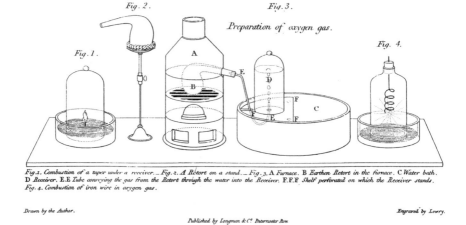

Fig. 1. Combustion of a taper under a receiver. — Fig. 2. A Retort on a stand. — Fig. 3. A Furnace. B Earthen Retort in the furnace. C Water bath. D Receiver. E.E Tube conveying the gas from the Retort through the water into the Receiver. F.F.F Shelf perforated on which the Receiver stands. Fig. 4. Combustion of iron wire in oxygen gas.

Drawn by the Author.　　　　　　　　　　　　　　　　　　　　　　　　　　　　　　*Engraved by Lowry.*

Published by Longman & Cᵒ Paternoster Row

Fig. 6.4 Illustration from *Conversations on Chemistry* concerning the preparation of oxygen and combustions (5th English edition, 1817, p. 181)

immediately embrace atomism. Nevertheless, as this chemist gradually accepted a corpuscular view of matter to some extent, subsequent editions of *Conversations* began including the concept of the atom [51, 52].

In Jane Marcet's book, experimentation plays a crucial role, and the way the various experiments are described creates a strong sense of realism. The descriptions are so vivid that readers can almost imagine themselves witnessing the experiments firsthand. Several of the experiments are accompanied by illustrations with the respective assembly schemes, which the author herself has drawn (Fig. 6.4). The following dialogue, about an experiment designed to demonstrate the concept of *attraction* between substances (which at the time was thought to be responsible for chemical combination), shows the clarity, detail, and liveliness of the descriptions:

Mrs. B.:　　[…] If, for instance, I pour on the piece of copper, contained in this glass, some of this liquid (which is called nitric acid), for which it has a strong attraction, every particle of the copper will combine with a particle of acid, and together they will form a new body, totally different from either the copper or the acid. Do you observe the internal commotion that already begins to take place?

Emily:　　The acid […] appears to be very rapidly dissolving the copper.

Mrs B.:　　By this means it reduces the copper into more minute parts than could possibly be done by any mechanical power. But as the acid can act only on the surface of the metal, it will be some time before the union of these two bodies will be completed.

You may, however, already see how totally different this compound is from either of its ingredients. It is neither colorless, like the acid, nor hard, heavy, and yellow like the copper. If you tasted it, you would no longer perceive the sourness of the acid. It has at present the appearance of a blue liquid; but when the union is completed, and the water with which the acid is diluted is evaporated, the compound will assume the form of regular crystals, of a fine blue color, and perfectly transparent. Of these I can show you a specimen, as I have prepared some for that purpose.

Caroline: How very beautiful they are, in color, form, and transparency!

Emily: Nothing can be more striking than this example of chemical attraction! [53]

On one hand, Jane Marcet's didactic approach values experimentation, and, on the other, her dialogues present a consistent pattern: they are structured to match the topic's development, ensuring an appropriate length. As a general practice, Mrs. B. introduces concepts by utilizing familiar examples and analogies, gradually presenting new ideas. Ultimately, she guides her students in drawing comparisons and deriving conclusions. Typically, each dialogue concludes with a recommendation to review the topic before proceeding to the next one [54]. In the upcoming dialogue, the author utilizes the analogy of bread to initiate a discussion on the concept of elementary substance:

Mrs. B.: [...] you must observe that the various bodies in Nature are composed of certain elementary principles, which are not very numerous.

Caroline: Yes; I know that all bodies are composed of fire, air, earth, and water; I learnt that many years ago.

Mrs B.: But you must now endeavour to forget it. I have already informed you what a great change chemistry has undergone since it has become a regular science. [...] an elementary body is one that has never been decomposed, that is to say, separated into other substances; and fire, air, earth, and water, are all of them susceptible of decomposition.

Emily: I thought that decomposing a body was dividing it into its minutest parts. And if so, I do not understand why an elementary substance is not capable of being decomposed, as well as any other.

Mrs. B.: You have misconceived the idea of *decomposition*; it is very different from mere *division*. The latter simply reduces a body into parts, but the former separates it into the various ingredients, or materials, of which it is composed. If we were to take a loaf of bread, and separate the several ingredients of which it is made, the flour, the yeast, the salt, and the water, it would be very different from cutting or crumbling the loaf into pieces.

Emily: I understand you now very well.

Caroline: But flour, water, and other materials of bread, according to our definition, are not elementary substances?

Ms B.:   No, my dear; I mentioned bread rather as a familiar comparison, to illustrate the idea, than as an example. The elementary substances of which a body is composed are called the *constituent* parts of that body; in decomposing it, therefore, we separate its constituent parts. If, on the contrary, we divide a body by chopping it to pieces, or even by grinding or pounding it to the finest powder, each of these small particles will still consist of a portion of the several constituent parts of the whole body: these are called the *integrant* parts; do you understand the difference?

Emily: Yes, I think, perfectly. We *decompose* a body into its *constituent* parts; and *divide* it into its *integrant* parts.

Mrs. B.: Exactly so. If therefore a body consists of only one kind of substance, though it may be divided into its integrant parts, it is not possible to decompose it. Such bodies are therefore called *simple* or *elementary*, as they are the elements of which all other bodies are composed. *Compound bodies* are such as consist of more than one of these elementary principles [55].

Conversations enjoyed enormous popularity, undergoing 16 editions in the United Kingdom between 1806 and 1852 [44]. It was translated into several languages, including Portuguese, as will be discussed. In the United States of America, where it had its own editions, its impact was equally remarkable: it was printed 23 times from 1806 to 1850 [56]. Thomas Jefferson, the third American president and a great enthusiast of chemistry, recommended it to those interested in learning this science [57]. Jane Marcet consistently updated the successive editions to align with the scientific developments of the time. In this regard, the opening note included in the fifth English edition is particularly noteworthy:

The author, in this fifth edition, has endeavored to give an account of the principal discoveries which have been made within the last four years in chemical science, and of the various important applications, such as the gas-lights, and the miner's-lamp, to which they have given rise.

Later, in the eighth edition, she introduced the recent discoveries in electro-magnetism and Proust's law of definite proportions [41]. This ongoing dedication to updating *Conversations* is also evident in the letter she wrote to Michael Faraday (1791–1867) at the age of 71:

Dear Mr. Faraday, I have this morning read in the "Athenaeum", some account of a discovery you announce … respecting the identity of the imponderable agents, heat, light, and electricity; and as I am at this moment correcting the sheets of my "Conversations on Chemistry" for a new edition, might I take the liberty of begging you to inform me where I could obtain a current account of this discovery? It is, I fear, of too abstruse a nature to be adapted to my young pupils; yet I cannot make up my mind to publish a new edition without making mention of it; I have, therefore, kept back the proof sheets of the Conversations on Electricity which I was this morning revising, until I receive your answer, in hopes of being able to introduce it in that sheet [58].

In 1806, the year *Conversations* was first published, Faraday was still a teenager, and it was through it that he received his initial chemistry lessons [59]. When the author passed away in 1858, the great physicist and chemist wrote a letter to the Swiss physicist Auguste de la Rive, who was closely associated with the Marcet family:

Mrs. Marcet was a good friend to me, as she must have been to many of the human race. I entered the shop of a bookseller and bookbinder at the age of 13, in the year 1804, remained there 8 years, and during the chief part of the time bound books. Now it was in these books, in the hours after work, that I found the beginnings of my philosophy. There were two that especially helped me; the *Encyclopedia Britannica*, from which I gained my first notions of Electricity and Mrs. Marcet's *Conversations on Chemistry*, which gave me my foundation in that science.

Do not suppose that I was a very deep thinker, or was marked as a precocious person. I was a very lively, imaginative person, and could believe in the Arabian Nights as easily as the Encyclopedia. But facts were important to me and saved me… so when I questioned Mrs. Marcet's book by such little experiments as I could find means to perform, and found it true to the facts as I could understand them, I felt that I had got hold of an anchor in chemical knowledge and clung fast to it. Hence my deep veneration for Mrs. Marcet… as one able to convey the truths and principles of those boundless fields of knowledge which concern natural things to the young, untaught, and inquiring mind [60].

The pioneering spirit and modernity of *Conversations on Chemistry*, which had already been translated into French in 1809, are particularly evident when compared to another popular science-themed work, also aimed at a female audience, published in Paris in 1810: *Lettres à Sophie Sur la Physique, la Chimie et l'Histoire Naturelle* [*Letters to Sophie on Physics, Chemistry, and Natural History*], written by Louis Aimé-Martin. Rather than aiming to provide real scientific instruction to its female readers, this work belonged to a literary genre inherited from the Enlightenment. Its purpose was to offer young women a series of scientific notions that would enable them to engage in conversations within the intellectual salons of society. Historian Natalie Pigeard refers to these works of scientific dissemination as "salon literature" and considers Aimé-Martin's work to be almost a caricature [61]. In fact, it predominantly contains stories and poems filled with classical and mythological references, along with expressions of praise and gratitude to divine providence for the perfection of Nature. It should be acknowledged, however, that veneration for divine creation is also present in *Conversations on Chemistry*, and it is with words of appreciation to the Most High that the author concludes her work:

> To God alone man owes the admirable faculties which enable him to improve and modify the productions of Nature, […] every acquisition of knowledge will prove a lesson of piety and virtue [62].

Despite the fact that *Conversations on Chemistry* does not utilize equations, formulas, or chemical symbols, it employs a genuinely scientific language for its time and includes a variety of experiments that could be conducted at home. To facilitate the comparison between Marcet's and Aimé-Martin's works, let us focus on the concept of *chemical attraction*, starting with *Conversations*:

Mrs. B.:   The term *attraction* has been lately introduced into chemistry as a substitute for the word *affinity*, to which some chemists have objected, because it originated in the vague notion that chemical combinations depended upon a certain resemblance, or relationship, between particles that are disposed to unite; and this idea is not only imperfect, but erroneous, as it is generally particles of the most dissimilar nature, that have the greatest tendency to combine.

Caroline: Besides, there seems to be no advantage in using a variety of terms to express the same meaning; on the contrary it creates confusion; and as we are well acquainted with the term Attraction in natural philosophy, we had better adopt it in chemistry likewise.

Mrs. B.:   If you have a clear idea of the meaning, I shall leave you at liberty to express it in the terms you prefer. For myself, I confess that I think the word Attraction best suited to the general law that unites the integrant particles of bodies; and Affinity better adapted to that which combines the constituent particles, as it may convey an idea of the preference which some bodies have for others, which the term *attraction of composition* does not so well express.

Emily: So I think; for though that preference may not result from any relationship, or similitude, between the particles (as you say was once supposed), yet, as it really exists, it ought to be expressed.

Mrs. B.:   Well, let it be agreed that you may use the terms *affinity, chemical attraction* and *attraction of composition*, indifferently, provided you recollect that they have all the same meaning.

Emily: I do not conceive how bodies can be decomposed by chemical attraction. That this power should be the means of composing them, is very obvious; but that it should, at the same time, produce exactly the contrary effect, appears to me very singular.

Mrs. B.:   To decompose a body is, you know, to separate its constituent parts, which, as we have just observed, cannot be done by mechanical means.

Emily: No: because mechanical means separate only the integrant particles; they act merely against the attraction of cohesion, and only divide a compound into smaller parts.

Mrs. B.:   The decomposition of a body is performed by chemical powers. If you present to a body composed of two principles, a third, which has a greater affinity for one of them than the two first have for each other, it will be decomposed, that is, its two principles will be separated by means of the third body. Let us call two ingredients, of which the body is composed, A and B. If we present to it another ingredient C, which has a greater affinity for B than that which unites A and B, it necessarily follows that B will quit A to combine with C. The new ingredient, therefore, has effected a decomposition of the original body AB; A has been left alone, and a new compound, BC, has been formed [63].

In turn, in *Lettres à Sophie*, no doubt in tune with the German writer Johann Wolfgang von Goethe (1749–1832), Aimé-Martin presents the following narrative:

The phenomena of attraction have given rise to some laws which govern the empire of chemistry. Those of men are much more numerous and they still live in war. Only one would suffice for them to be happy: to possess the power of loving.

As the laws which have been deduced from the chemical action of bodies are very complicated, I will content myself, for the present, with expounding to you one of them which it is indispensable to know. Do not laugh at my modest scientific tone, and do not ask me to hide anything from you. My only aim is to avoid difficulties for you. I cannot, for the present, offer you more than the flowers of science: but remember that the first flowers with which spring crowns itself are those that promise delicious fruit.

The law of which I wish to speak to you is known by the name of elective attraction; it is, if I may so express it, the love which leads one of the substances of a compound to abandon the body of which it forms a part, in order to unite itself with a new substance, its preferred one.

A surprising phenomenon, which seems to establish a certain friendship between the most insensible bodies!

If the ancients, who fantasized everything, had known of these mysteries, they would have created a multitude of laughing nymphs, who, yielding to the impetus of the heart, would have preserved in their metamorphoses the sweet inclination to impermanence; Ovid would have sung them, holding Cupid's lyre to him. […]

It is to the elective affinities that we owe the harmony that reigns in the elements of the world, as well as the constant reproduction of flowers, fruit, air and water. If one substance were not destined to unite with a given substance instead of another, everything would revert to chaos, everything would be confused; or, to put it better, nothing that exists would exist. The world would be no more than a mass of these simple bodies, these primitive elements, of which we know only a part.

Chemistry is, therefore, nothing more than the art of discovering, seconding and imitating these diverse affinities. But how imperfect are the operations of science in comparison with those of Nature! [64]

The existence of at least 10 editions of this work suggests that the French public had a clear preference for a more literary approach to scientific dissemination [65].

Although indirectly, *Conversations on Chemistry* was also translated and adapted into Portuguese, having been published in 1834 under the title *A Chimica Ensinada em 26 Liçoens* [*Chemistry Taught in 26 Lessons*] [66]. This has been done by António Teixeira Girão, who used as a source the French adaptation that the chemist Anselme Payen had made of Marcet's work [67]. In this version, published in Paris and Brussels in 1825, Payen departed freely from *Conversations on Chemistry* and made several alterations. Among these changes, he dropped the dialogue between the governess and the pupils, which, in itself, substantially shortened the text despite the introduction of additional information. As a result, the version was condensed into a single volume of 442 pages. One spurious content in this edition is Dalton's atomic theory [68]. On the other hand, Teixeira Girão's translation is an expanded version of Payen's, spanning 516 pages. Peres and Rodrigues suggest that Girão was unaware that the original work had been written by a woman [69].

To conclude, Genevan Albertine Necker de Saussure (mentioned in Chap. 5) acknowledged that Jane Marcet's book had a significant influence on her writing of *L'Éducation Progressive* [*Progressive Education*, 1828], a work that brought her fame. In this book, she advocated for the inclusion of science, especially chemistry, as a vital component of female education [70].

6.4   Mrs. B

While it is possible that the characters in *Conversations on Chemistry* were not directly inspired by real-life individuals, this has not hindered the formulation and ongoing consideration of various hypotheses. The most usual one suggests that Margaret Bryan (fl. 1795–1816), an educator and science popularizer specializing in physics and astronomy, served as the model for Mrs. B., while her two daughters, whose names are unknown, were the inspiration for the characters Emily and Caroline (Fig. 6.5).

In truth, Jane Marcet's correspondence reveals that as early as 1803, she had already selected the names for the two children in her book. This makes it impossible to imagine that they were inspired by her own children, as has also been suggested, since her first daughter was not yet born. In 2016, two historians, Leigh and Rocke, offered a different explanation: the names belonged to Emily and Caroline Sebright, daughters of Sir John Sebright, a member of the Marcet circle [71]. What follows is based entirely on the conclusions of these two researchers.

Sir John Sebright, described as an eccentric and at times disagreeable figure, had a notable reputation as an amateur chemist and maintained close friendships with renowned individuals like William Wollaston and Humphry Davy. Alongside his chemical pursuits, Sir John also had a keen interest in breeding dogs, hawks, pigeons, and chickens. He wrote three monographs on the subject, which Charles Darwin referenced in his work *The Variation of Animals and Plants under Domestication* (1859). Sir John had eight daughters, reportedly described as ungraceful and even unattractive by some accounts, along with one son. During the spring of 1803, when the names "Emily" and "Caroline" first appeared in the manuscript of *Conversations on Chemistry*,

Fig. 6.5 Margaret Bryan with her daughters in the frontispiece of her *A Compendious System of Astronomy* (1797). Engraving by William Nutter. National Portrait Gallery, London

Emily and Caroline Sebright were, respectively, 6 and 5 years old. Although they were too young to study chemistry, as suggested by Leigh and Rocke, they could have potentially inspired Jane Marcet. Sir John, in any case, took great care in educating his daughters, personally providing them with dance lessons accompanied by his violin and teaching them chemistry. Situated approximately 40 miles north of London, their family estate, Beechwood, boasted a laboratory. This is not surprising, as chemistry was highly valued during that time for its practical applications in industry and agriculture, in addition to being a pastime for the wealthy.

In London, Davy's captivating lectures at the Royal Institution were the talk of the town, and he himself championed the education of women in the field of chemistry. Thus, it is not impossible that as early as 1803, Sir John had intentions of providing his daughters with an education in this scientific domain, and the Marcets may have been aware of his plans. Jane may have then seen in the real Emily and Caroline the models for the young characters in her book. Frederica, the eldest daughter of Sir John, was already 7 years old in 1803. However, since her name was not commonly used for an English girl, the author of *Conversations* chose to use her sisters' names instead. Nevertheless, this hypothesis has a significant weakness, as acknowledged by Leigh and Rocke: the two girls are not portrayed as sisters in Jane Marcet's book: the two girls are not depicted as sisters in Jane Marcet's book. Regardless, the most significant finding of these two historians' research was the revelation of a remarkable chemist from the early nineteenth century, now largely forgotten: Frederica Sebright herself!

In the summer of 1812, the young yet already renowned Swedish scientist Jöns Jacob Berzelius visited England and spent considerable time with the Marcets. He also stayed a few days with the Sebright family at Beechwood. In his diary entry from that period, he wrote:

> Sir John showed me his laboratory, which was very nicely set up, and equipped with everything necessary for chemical research. [...] I passed a pleasant and interesting week there in the company of Sebright's family and the chemists [Alexander] Marcet and [Richard] Chenevix [and their wives]. [...] Sebright had built a quite clever chemical laboratory on his estate for his daughters [72].

The analysis of the correspondence between Alexander Marcet and Berzelius led to the conclusion that Berzelius had engaged in discussions about chemistry with Sir John's daughters. At that time, Frederica was 16, Emily was 15, and Caroline was 14 years old. It also indicated that Berzelius had visited the laboratory with them. In a letter written 3 months later, Berzelius asks his correspondent to convey his regards to all his friends, especially the *chemists* ("*the chemists of the fair sex*"), likely referring to the Sebright sisters. In another letter, written 3 years later, Alexander Marcet mentions Frederica's exceptional chemistry skills. Toward the end of 1815, Berzelius writes to Alexander Marcet, sharing the information he received about two young English girls seen in the laboratory of the renowned French chemist and pharmacist Louis-Nicolas Vauquelin (1763–1829). Based on the description provided, he believed that these girls were none other than Frederica and Emily Sebright:

You ask me if it really was the Misses Sebright who were seen in Vauquelin's laboratory. It was them, and I assure you that they had every right to be in that sanctuary. Certainly the elder is really a very good chemist. The other day she pointed out to Wollaston certain brown particles mixed with grains of native iridium, and she asked him what these particles were. He didn't know, and suggested she investigate them herself. She set to work and soon said that they were composed of iron, manganese, and tungsten. Which Wollaston found to be completely correct.

Wollaston himself narrated this episode to Berzelius:

I believe it was in your letter to Marcet that you mention our friends at Beechwood. You will be gratified to hear that the eldest Miss Sebright is really becoming an expert analytical chemist with considerable skill and ingenuity in operating upon microscopic quantities [72].

The fact that Frederica and Emily visited Vauquelin's laboratory, when they would have been 19 and 18 years old, respectively, suggests that they may have embarked on a trip to continental Europe in 1815, shortly after Napoleon's final defeat at the Battle of Waterloo. It is highly probable that they carried letters of recommendation from Alexander Marcet and Wollaston, as Vauquelin was known for welcoming chemists who were interested in working in his laboratory at the Natural History Museum in Paris.

There are additional accounts of Frederica Sebright's accomplishments in the field of chemistry. One such testimony comes from Archdukes Johann and Ludwig of Austria, who visited Beechwood during their trip to England in 1815. "We passed the evening very agreeably," they wrote, adding that

the baronet's eldest daughter, who devotes much of her time to the study of chemistry, showed us an experiment of Wollaston's [1812 or 1813], which is now known, but was then new to us, and which consists in transforming a thimble into a small galvanic battery capable of heating a platinum wire red hot.

Dated 1817, there is also this statement from Marcet to Berzelius:

Sir John is still a *grand amateur*, but his eldest daughter [Frederica was now age 21] is a real chemist. She works enthusiastically and analyses very successfully. She has just determined the weight of the atom of phosphorus by a new method which is in agreement with the best authorities, and particularly with you.

Half a year later, the Swedish chemist returned to London, but had no opportunity to meet with the Sebright family. In a letter to Marcet, he send:

A thousand greetings to [various friends] and the same to M. Sebright and to those young ladies, and I beg you to tell the chemist [*la chimiste*] how sorry I am not to have had the opportunity to tell her that I am delighted by the asset that chemistry has acquired in her person [72].

The praises from chemists such as Wollaston and Berzelius serve as strong evidence of Frederica Sebright's exceptional abilities in chemical analysis. Emily, on the other hand, tragically passed away just 7 months after her marriage to a young naval officer in 1822, the cause of her death remaining unknown. However, her husband found love once again, and in 1824, he married Frederica, his late wife's sister, in New York. The couple enjoyed a happy marriage and had six children, with their eldest son being named Augustus Wollaston in honor of his godfather, William Hyde Wollaston. Frederica passed away in London in 1864 at the age of 68.

6.5 Chemistry for Everyone

In the nineteenth century, Almira Hart Lincoln Phelps (1793–1884) emerged as a notable figure in the United States as an author of science textbooks targeted at women (Fig. 6.6). Born in Berlin, Connecticut, her first teacher was her sister, Emma Hart Willard (1787–1870), who not only worked as an educator but also advocated for women's rights, as explored in Chap. 1 of the next volume. Almira taught in various rural schools and academies in Connecticut and New York. In 1824, after the death of her first husband, she began teaching languages and sciences at the Female Seminary founded by her sister 3 years earlier in Troy, New York.

During this time, opportunities for women to pursue careers in science were extremely limited. However, at the newly established Rensselaer Institute, also located in Troy, there was a distinguished faculty member named Amos Eaton (1776–1842) who supported the idea of girls studying and encouraged them to attend his theoretical and practical courses [73, 74]. The two Hart sisters became his students, and Almira, in particular, received a strong scientific education. In 1831, following her second marriage and adopting the surname Phelps, Almira was encouraged by her husband to pursue her talents as a didactic writer. Subsequently, she published books in various fields, including botany, geology, physics, and chemistry. In the realm of chemistry,

Fig. 6.6   Almira Hart Lincoln Phelps, *c.*1853

Almira gained recognition in 1830 for her translation of a French dictionary. She further distinguished herself with the textbooks *Chemistry for Beginners* (1834) and *Familiar Lectures on Chemistry* (1837) [75, 76]. In the preface to the former, one can read:

> The author has sought to present the elements of the science in a popular and attractive form, without offending the scholar by marring its classical beauty and proportions [77].

A little further on, she gives the following account:

> A scientific professor once said to a lady in a sarcastic tone – "and so *you* are going to attempt to teach Chemistry?" "Yes", said she, good-naturedly, "as fast as I can learn it myself". The same professor was afterwards engaged to lecture, and perform experiments before the lady's class, but her pupils complained that his language was so little adapted to their attainments, that they could not understand his lectures until explained to them in the familiar language of the teacher [77].

Then, in the introduction to the first chapter ("Of the nature and importance of chemistry"), while emphasizing the merits of applied chemistry, she mentions various household activities:

> This science bears an important relation to housekeeping in a variety of ways, as in the making of gravies, soups, jellies and preserves, bread, butter and cheese, in the washing of cloths, making soap, and the economy of heat in cooking, and warming rooms: – To females, then, some knowledge of chemistry must be very desirable [78].

She adds also that young women who study chemistry:

> [...] should therefore pay strict attention to all those facts in housekeeping which may be explained upon chemical principles, such as the action of yeast upon flour, and of pearl ash [potassium carbonate] upon sour dough, the change of cider into vinegar, the advantage of keeping a vessel covered in order to hasten the boiling of water, etc. [78].

Like Jane Marcet's *Conversations*, Almira Phelps' chemistry textbooks feature illustrations depicting laboratory equipment and experimental setups created by the author herself. In a note directed to teachers, immediately following the preface to *Chemistry for Beginners*, she makes the following observation:

> It is desirable that teachers should be able to make some experiments, let them be ever so simple; but some who use this book, will probably be so situated as to render it very difficult. [...] In view of this, I have endeavored to make the work plain and simple, and to give such drawings as might compensate, as far as possible, for the want of experiments. [...] In the Female Seminary, at Troy, where for several years, I had the charge of the chemical department, the pupils were required to perform experiments, and to give lectures before the class on given subjects [79].

In the 1834 edition, within chapter XI, which is dedicated to hydrogen, the author reiterates the importance of experimentation in a footnote:

> The young ladies of the Seminary of Troy N. Y. who are in the habit of performing chemical experiments in their daily exercises and at the public examinations, have by means of a suitable apparatus, exhibited some splendid experiments, to illustrate the burning of hydrogen and carburetted hydrogen [methane] [80].

Properties of Iodine.

Fig. 6.7   Illustration from *Chemistry for Beginners* (1838 edition, p. 134)

In an edition of 1838, an illustration (Fig. 6.7) depicts a girl actively engaging in an experimental study of the properties of iodine—and seemingly inhaling it!

The poem "The Chemical Conviction" at the beginning of this chapter is part of a collection of over 200 poems by Emily Dickinson (1830–1886) that have connections to the natural sciences, medicine, and technology. It is known that her secondary studies included chemistry. While attending Mount Holyoke Female Seminary in Massachusetts for two academic terms, she wrote to her brother Austin in January 1848, expressing that his last letter had found her "all absorbed in the history of sulfuric acid!!!!!" (exactly with five exclamation marks!) [81]. Emily Dickinson, who studied botany through the very popular *Familiar Lectures on Botany* (1829) by Almira Phelps, would most likely have been familiar with *Chemistry for Beginners* as well. Nevertheless, the textbook she mentions in her correspondence is the recently released *First Principles of Chemistry* (1847) by Benjamin Silliman Jr., which had been adopted at Mount Holyoke [82, 83]. It is interesting to note that the formula for sulfuric acid in this work is given as $SO_3.HO$ (and for water as HO) [84], which aligns with the information found in Phelps's *Chemistry for Beginners*. According to Phelps, sulfuric acid consists of "one atom of sulfur = 16, united

Fig. 6.8 Mary Fairfax Somerville. Library of Congress Prints and Photographs Division Washington, D.C.

with three atoms of oxygen = 24 [85]". It should be acknowledged that these atomic masses correspond to the values proposed by Justus von Liebig (the correct masses for these elements are approximately double of what is stated: 32 and 16, respectively) [86].

The Scottish mathematician and science communicator Mary Fairfax Somerville (1780–1872) also warrants recognition in this context due to her exceptional talent for presenting and synthesizing scientific concepts, establishing her reputation as a highly accomplished science communicator (Fig. 6.8). Despite not being a scientist herself, she achieved immense international recognition and was even hailed as the "Queen of Nineteenth-Century Science."

As the daughter of a naval officer, Mary received a very basic education, unlike her two brothers. Her father, who frequently spent long periods at sea, was shocked to discover that Mary, at nearly 9 years old, could barely read and write. In response, he promptly enrolled her in a boarding school where she received her only formal instruction throughout her life, lasting for just 1 year. It was during this time that Mary's latent passion for mathematics began to emerge. A few years later, her previously undiscovered inclination toward mathematics was serendipitously revealed. Mary came across an algebra problem in a women's magazine and, with the aid of a book on navigation, successfully solved it. It wasn't until she turned 15 that she managed to get her hands on an introductory mathematics textbook, along with a copy of Euclid's *Elements of Geometry*, which she read and interpreted by herself, keeping it hidden from her parents. When her father discovered her secret, he forbade her from reading such books, fearing that the intellectual exertion might drive her to madness—after all, one never knew what could happen with girls!

Until 1804, when she married a naval officer, she spent winters in Edinburgh, where she indulged in parties, balls, and theater. She enjoyed reading poetry, playing the piano, painting, embroidering, and teaching herself French, as well as some Greek and Latin. During the summers in Burntisland, she spent extended hours observing Nature, which sparked her interest in natural history. After her marriage, she found more time for reading and learning, despite her husband's disapproval of women pursuing education. His untimely death in 1807 granted Mary economic independence and freedom to fully immerse herself in the study of mathematics. In Edinburgh, she discovered an intellectual environment that nurtured her scientific ambitions. In 1812, she entered into a second marriage with William Somerville, a military physician and her cousin, who proved to be supportive and understanding throughout their 50-year marriage, encouraging her scientific pursuits. In 1816, they relocated to London, where they quickly became acquainted with the leading intellectual circles and established connections with notable figures such as Alexander and Jane Marcet, Humphry Davy, Thomas Young, and William Wollaston. During their visit to Paris in 1817 (as mentioned in the previous chapter), Mary and William Somerville had the privilege of meeting the renowned French mathematician and astronomer Pierre-Simon Laplace. He would later acknowledge Mary Somerville as the "only woman in England who could understand and correct his works" [87]. In 1826, Mary achieved a significant milestone by publishing her first paper in *Philosophical Transactions*, the journal of the Royal Society. The paper focused on the purported magnetizing effect of sunlight. However, it was her four books written over a span of nearly four decades that would bring her widespread acclaim. The first one, *The Mechanism of the Heavens*, was published in 1831 when she was already 51 years old. More than just a translation of Laplace's celestial mechanics, it also served as a textbook adopted for advanced mathematics courses at Cambridge University. This was followed by *On the Connexion of the Physical Sciences* in 1834, *Physical Geography* in 1848, and, at the remarkable age of 89, *On Molecular and Microscopic Science* in 1869, the same year that Dmitri Mendeleev arranged the chemical elements in the Periodic Table. It is this later work that places Mary Somerville within the context of this chapter, as it delves into various subjects, including the exploration of the atomic and molecular composition of matter. In the preface of her book, she tells us:

> Microscopic investigation of organic and inorganic matter is so peculiarly characteristic of the actual state of science, that the author has ventured to give a sketch of the most prominent discoveries in the life and structure of the lower vegetable and marine animals in addition to a few of those regarding inert matter [88].

The first part of the first volume contains the following sections: Elementary Constitution of Matter, Forces and Their Relations to Matter, Atomic Theory, Analysis and Synthesis of Matter, Utilization of Waste Substances (such as coal tar dyes), The Solar Spectrum, Spectral Analysis, Spectrum of Gases and Volatilized Matter, Color Line Inversion, and the Constitution of the Sun and Stars. The second part of the first volume focuses on organisms of plant origin, while the second volume explores organisms of animal origin. It is worth noting that the author still employs Liebig's atomic masses ($H = 1$, $C = 6$, $O = 8$) rather than Berzelius's values, which were already close to the correct values [89]. Notably, the work delves into contemporary subjects including spectroscopy, which was a relatively recent field at the time as the spectroscope had been invented in 1859 by Gustav Robert Kirchhoff and Robert Bunsen. The book also explores August von Hofmann's aniline dyes (synthesized from compounds present in coal tar, which was a byproduct of illuminating gas production by heating that form of coal), as well as mauveine, a purple aniline dye synthesized by William Perkin in 1856.

More than a popularizer of science, Mary Somerville was an expositor of science. In all of her works, she aimed to present the state of science at the time, even if she had to lay the foundations and provide definitions necessary for a reasonably educated audience to comprehend [90]. Mary Somerville passed away in Naples at the age of nearly 92. The latter part of her life was spent in Italy, where she and her husband had relocated in the late 1830s for his health reasons. Two years after her death, *Nature* magazine proclaimed her to be an exception to the belief that "women are not by nature adapted for studies which involve the higher processes of induction and analysis" [91]. Furthermore, it is worth noting that the term "scientist" was coined in 1834 by the mathematician and astronomer William Whewell in a critical review of her *On the Connexion of the Sciences*.

During Victorian England, there was a strong enthusiasm for popular science among the general public. People avidly collected items like ferns, shells, and minerals. They also enjoyed activities such as maintaining aquariums and terrariums, stargazing, and conducting chemical experiments at home. It this context, a series of children's books emerged that blended science and entertainment. One notable example is *The Fairy-Land of Science* (1879). Its author, Arabella Burton Buckley (1840–1929) was an important communicator of science to children, particularly in the field of Darwinian evolutionism. She had served for many years as the secretary to Charles Lyell (1797–1785), the famous British geologist [92]. Buckley believed that children could learn science through the use of stories and fables, provided that the "object lesson" format was employed. At the time, this teaching method had reached its peak during the lectures given by Michael Faraday at the Royal Institution, which eventually led to the publication of *The Chemical History of a Candle* in 1861. Similarly, Thomas Huxley's lectures, in which he narrated the geological history of Great Britain based on a piece of chalk, were published in 1868 under the title *On a Piece of Chalk* [93].

In the footsteps of writers such as Charles Kingsley (1819–1875), who, through works like *Glaucus, or The Wonders of the Shore* (1855), had been one of the initiators of the Christian-inspired enthusiasm that the Victorian public displayed for knowledge of the natural world, Buckley also aimed to communicate science while distancing it from mechanistic and materialistic theories. Instead, she promoted it in moralistic terms. Buckley believed that learning served not only as a means for individuals to acquire knowledge but also as a way to better themselves [94]. In *The Fairy-Land of Science*, the author conveys to her readers the notion that science narrates tales of an enchanted natural world full of stunning imagery, authentic poetry, and enchanting fairies. These fairies, in fact, represent invisible natural forces or phenomena such as heat, gravitation, or crystallization [95].

Along the same lines as Arabella Buckley, the American writer Lucy Rider Meyer (1849–1922) published in 1887 *Real Fairy Folks*, a book with the subtitle *Fairy Land of Chemistry: Explorations in the World of Atoms* (Fig. 6.9). Rider Meyer was an educator and physician with a background in chemistry. After having studied at the Massachusetts Institute of Technology, she taught chemistry for 2 years at McKendree College in Lebanon, Illinois. In her book *Real Fairy Folks*, Buckley utilizes a fairy tale approach to illustrate chemical concepts, portraying molecules in the form of fairies. The water molecule, for example, is represented by depicting the *oxygen fairy* holding hands with *two hydrogen fairies* (Fig. 6.9) [96].

a   b

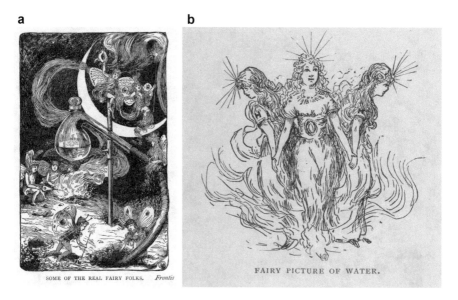

Fig. 6.9 (a) Frontispiece of Lucy Rider Meyer's *Fairy Land of Chemistry: Explorations in the World of Atoms*, Lothrop Publishing Company, Boston, 1887. (b) Illustration of the water molecule, p. 208. Courtesy of the Science History Institute

Just before the end of the century, Lily Martin, an English author, published a book called *The Princess & Fairy* (1899). With the subtitle *The Wonders of Nature*, it aims to teach children about various facts and phenomena of the natural world. This is achieved by its two characters, a Princess and a Fairy. The book covers many topics, from living organisms to matter, including the physical states and chemical elements. In one illustration, Fig. 6.10, the Princess tries to make *holes* in water in a basin. The Fairy explains that there are forces that hold the water's tiny particles together. The Princess asks if these particles are the drops, but the word "molecule" is never used [97].

There were few women with careers initiated before 1900 whose contributions to technical-scientific written works were significant. One remarkable exception was Martha Annie Whiteley (1866–1956), an Englishwoman who became the first female professor of chemistry at Imperial College London, a topic that will be explored in Chap. 3 of the next volume. She had a long and fruitful collaboration with the *Dictionary of Applied Chemistry*, which was initiated in 1890 by the English chemist Thomas Edward Thorpe and aimed to cover various chemistry-related topics. Initially serving as a coeditor, Martha Whiteley eventually became the editor-in-chief starting from the fourth edition in 1940. When she took part in the last volume in 1954, she was already 88 years old [98].

The Princess tries to make Holes in the Water.

Fig. 6.10 Illustration from *Princess & Fairy or The Wonders of Nature*. Courtesy of the Science History Institute

However, the female figure who contributed most to the professional literature of chemistry was the American Mary Fieser (1909–1997), even though much of her work was done in collaboration with her husband, the chemist Louis Fieser (1899–1977). Mary, née Peters, was born in Atchison, Kansas in 1909 into a cultured family. Her father was a college professor of English and her mother had attended college. Mary studied chemistry at Bryn Mawr College in Pennsylvania, where she was a student of Fieser. Like the vast majority of other students, Mary found Fieser's classes especially motivating. His talent for highlighting the experimental aspects of chemistry was the secret behind his captivating influence on students.

In 1930, the year Mary finished her bachelor's degree (Fig. 6.11), Fieser was invited to join the faculty at Harvard University. He accepted, but he did not go alone: Mary had decided to accompany him. With the desire to pursue further studies, she decided to enroll at Radcliffe College, women's college associated with Harvard University. After classes, she dedicated her time to conducting research alongside Fieser [99]. However, the environment at Harvard was not particularly welcoming for female students, and Mary had to face hostility,

Fig. 6.11 Mary Peters in the 1930 yearbook. Courtesy of Special Collections, Bryn Mawr College Libraries

especially from Gregory P. Baxter, a renowned professor of analytical chemistry known for his work on atomic masses, who refused to grant her access to the institutional laboratories. Consequently, she was compelled to conduct her experiments in an abandoned basement. Despite these obstacles, by 1932, she had accumulated enough results to obtain her master's degree in organic chemistry. In that same year, she married Fieser and made the decision not to pursue a Ph.D. However, there were no barriers to continuing her ongoing research. As Mrs. Fieser, she stated, "she could do all the chemistry she wanted" [100]. She continued working within her husband's research group, affectionately referring to him as "chief", even though she was not formally employed or paid, even after acquiring the status of a researcher much later. Throughout the 1930s and 1940s, she served as Fieser's primary collaborator in studying the chemistry of quinones and other natural products, particularly steroids. Her research career extended until 1956, during which she authored over 40 scientific papers [100].

In 1942, the couple embarked on the task of writing organic chemistry textbooks, with Mary taking charge of the literature research. Louis, who juggled research, teaching, and supervising students, realized that he could not process information as quickly as his wife collected it. Thus, he proposed that Mary begin writing some chapters herself. The first fruit of their collaboration was *Organic Synthesis*, published in 1944, which garnered remarkable success in both the United States and Europe. Encouraged by this reception, they felt compelled to continue their writing venture. Subsequent titles included *Introduction to Organic Chemistry* (1946), *Basic Organic Chemistry* (1958), *Advanced Organic Chemistry* (1961), and *Topics in Organic Chemistry* (1963), among others. What set these textbooks apart from others was not only their inherent quality and clarity of writing, but also the meticulous selection of terms and the inclusion of concrete examples of experimentally verified chemical reactions, rather than just the typical theoretical sequences. Moreover, the Fieser couple's textbooks also stood out due to the quality of the diagrams and reaction schemes, which, by themselves, served as important sources of information. Additionally, the inclusion of footnotes featuring concise biographical details about the cited chemists added a historical and human touch to their works. The popularity of their books only began to wane in the 1960s, when the teaching of organic chemistry began to favor deeper theoretical approaches [100].

The Fiesers were also the authors of *Style Guide for Chemists* (1960), a manual that offered writing advice for the chemical community. It included rules for the proper use of the English language, as well as guidance on the correct pronunciation of chemical terms and the reporting of experimental procedures [101]. However, their most renowned achievement was the monumental publishing project called *Reagents for Organic Chemistry*. Initiated in 1967, it remains active to this day (the 29th volume was released in 2019). This series offers a comprehensive listing of all reagents mentioned in the organic synthesis literature, covering them from A to Z. Each entry provides concise descriptions, diagrams illustrating chemical reactions, and bibliographic references. To accomplish this, Mary, who was proficient in French and German, dedicated 15 hours a day to meticulously examining organic chemistry literature from around the world, including journals of limited circulation. Until Louis' death in 1977, six volumes were published, and Mary, renowned for her exceptional analytical and retention skills, continued the endeavor until 1994 when the 17th volume was released [102]. The tremendous success of *Reagents for Organic Chemistry* stems primarily from its ability to keep organic chemists up to date on new reagents and synthesis methods, thus sparing them extensive and time-consuming literature searches.

In 1971, Mary Fieser was awarded the Garvan Medal by the American Chemical Society in recognition of her significant contributions to the literature of organic chemistry. In 1996, Harvard University honored Mary and Louis Fieser by naming its renewed organic chemistry teaching laboratory after them. Mary passed away the following year at the age of 87 at her home in Belmont, Massachusetts [101]. In 2008, the Department of Chemistry and Chemical Biology at Harvard University established the "Mary Fieser Postdoctoral Fellowship" with the aim of supporting underrepresented women and minorities in various fields of chemistry [103].

Despite not having children, the Fiesers were known for their love of cats and often had up to seven as companions. The prefaces of their books frequently featured illustrations of their beloved feline pets, becoming a sort of trademark image. Their first cat, a Siamese, was named "Synthetic Vitamin K Pooh." The naming was inspired by the fact that around the time it was born, the Fiesers had successfully synthesized vitamin K_1 by reacting phytol with the reduced form of methylnaphthoquinone [100, 104].

References and Notes

1. Dena Goodman, "Enlightenment Salons: The Convergence of Female and Philosophic Ambitions", *Eighteenth-Century Studies*, 22 (1989) 329–350
2. M. W. Rossiter, "Women and the history of scientific communication", *Journal of Library History*, 21 (1986) 39–59
3. Marelene Rayner-Canham, Geoffrey Rayner-Canham, *Pioneering British Women Chemists: Their Lives and Contributions*, World Scientific, London, 2020, p. 12
4. Londa Schiebinger, *The Mind Has No Sex?: Women in the Origins of Modern Science*, Harvard University Press, Cambridge (MA), 1989, pp. 238–240
5. Jean-Jacques Rousseau, *Émile, ou de l'éducation (notices et annotations par Henri Legrand)*, Larousse, Paris, 1914, p. 199
6. Leigh Ann Whaley, *Women's History as Scientists – A Guide to the Debates*, ABC Clio, Santa Barbara, 2003, pp. 117–121
7. Schiebinger, *Op. cit.* (4), pp. 236–237
8. *Ibid.*, pp. 220–223
9. D. J. [Louis de Chevalier de Jaucourt], "Squelette", *Encyclopédie, ou Dictionnaire raisonné des sciences, des arts, et des métiers*, Neuchâtel, 1765, vol. 15, pp. 482–483
10. D. Williams, "The Fate of French Feminism: Boudier de Villemert's Ami des Femmes", *Eighteenth-Century Studies*, 14 (1980) 37–55
11. Schiebinger, *Op. cit.* (4), pp. 240–242
12. Jorge Calado, *Haja Luz – Uma História da Química Através de Tudo*, IST Press, Lisboa. 2011, pp. 30–31
13. A. B. Shteir, "Botanical Dialogues: Maria Jacson and Women's Popular Science Writing in England", *Eighteenth-Century Studies*, 23 (1990), 301–317
14. Rebecca M. Messbarger, *The Century of Women: Representations of Women in Eighteenth-century Italian Public Discourse*, University of Toronto Press, Toronto, 2002, pp. 73–75

15. P. Findlen, "Becoming a Scientist: Gender and Knowledge in Eighteenth-Century Italy", *Science in Context*, 16 (2003) 59–87

16. *Ouvres de Fontenelle – Études sur sa vie et son esprit*, Eugène Didier, Paris, 1852, p. 35

17. F. Lussana, "Misoginia e adulazione: ambiguita' dell'immagine femminile nel Secolo dei Lumi", *Studi Storici*, 25 (1984) 547–558

18. F. Arato "Minerva e Venere: Scienze e lettere nel Settecento italiano", *Belfagor*, 48 (1993) 569–584

19. Monique Frize, *Laura Bassi and Science in 18th Century Europe – The Extraordinary Life and Role of Italy's Pioneering Female Professor*, Springer, Heidelberg, 2013, pp. 1–6

20. Messbarger, *Op. cit.* (14), pp. 73–75

21. Laboratory device that measures the change in volume of a gas mixture after a physical or chemical change.

22. A. Elena, "In lode della filosofessa di Bologna": An Introduction to Laura Bassi, Isis, 82 (1991) 510–518

23. G. B. Logan, "The Desire to Contribute: An Eighteenth-Century Italian Woman of Science", *The American Historical Review*, 99 (1994) 785–812

24. F. Lussana, "Il concetto di uguaglianza e il dibattito sulla donna nella Francia prerivoluzionaria e in Italia", *Studi Storici*, 31 (1990) 437–455

25. Frize, *Op. cit.* (19), pp. 19–21

26. *Novelle della Repubblica delle lettere dell'anno MCCLVIII, pubblicate sotto gli auspizi dell' eminentiss. e reverendiss. Principe Carlo Rezzonico*, Venezia, 1758, pp. 129–131

27. Gregorio Piaia, Giovanni Santinello (Eds), *Models of the History of Philosophy: Vol. III: The Second Enlightenment and the Kantian Age*, Springer, Dordrecht, 2015 p. 320

28. Giusepe Compagnoni, *La Chimica Per le Donne*, Venezia, 1796, Tomo primo, p. v.

29. *Ibid.*, p. vi

30. *Ibid.*, p. viii

31. *Ibid.*, p. 1

32. *Ibid.* pp. 1–2

33. *Ibid.* pp. 5–6

34. *Ibid.* p. 6

35. P. Findlen, "Translating New Science: Women and the Circulation of Knowledge in Enlightenment Italy", *Configurations*, 3 (1995) 167–206

36. Carlo Luigi Benvenuto Robbio di San Raffaele, *Disgrazie di donna Urania ovvero degli studi femminili*, Stamperia Granducale, Firenze, 1798, pp. 26–28

37. Aaron J. Ihde, *The Development of Modern Chemistry*, Harper & Row, New York, 1964, cap. 4

38. M. Sutton, "A clash of symbols", *Chemistry World*, November 2008, 56–60

39. Ihde, *Op. cit.* (37), cap. 5

40. L. Rosenfeld, "The Chemical Work of Alexander and Jane Marcet", *Clinic. Chem.*, 47 (2001) 784–792

41. E. V. Armstrong, "Jane Marcet and her *Conversations on Chemistry*", *J. Chem. Educ.*, 15 (1938) 53–57

42. N. G. Coley, "Alexander Marcet (1770–1822), Physician and Animal Chemist", *Medical History*, 12 (1968) 394–402

43. A. De la Rive, "Madame Marcet", *Bibliothèque Universelle – Revue Suisse et Étrangère*, 64 (1859), pp. 454–455

44. G. J. Leigh, "The changing content of *Conversations on Chemistry* as a snapshot of the development of chemical science", *Bull. Hist. Chem.*, 42 (2017) 7–28

45. Jane Marcet, *Conversations on Chemistry*, vol. 1, 5th edition, Longman, London, 1817, p. v

46. A. B. Shteir, "Botanical Dialogues: Maria Jacson and Women's Popular Science Writing in England", *Eighteenth-Century Studies*, 23 (1990), 301–317

47. John Issitt, *The life and works of Jeremiah Joyce*, PhD thesis, The Open University, 2000, pp. 197–200, http://oro.open.ac.uk/58059/1/343513.pdf

48. Rayner-Canham, *Op. cit.* (3), p. 15

49. Marcet, *Op. cit.* (45), pp. vi–vii

50. M. Meyer, G. Patterson, "Mrs. Jane Marcet and «Conversations in Chemistry»", in Gary Patterson (ed.), *Preceptors in Chemistry*, ACS Symposium Series, vol. 1273, Washington DC, 2018

51. R. Siegfried, "The Chemical Philosophy of Humphry Davy", *Chymia*, 5 (1959) 193–201

52. P. Ball, "The marvellous Mrs Marcet", *Chemistry World*, 11 January 2018, https://www.chemistryworld.com/opinion/the-marvellous-mrs-marcet/3008457.article

53. Marcet, *Op. cit.* (45), pp. 17–19

54. M. E. Derrick, "What can a nineteenth century chemistry textbook teach twentieth century chemists?", *J. Chem. Educ.,* 62 (1985) 779–751

55. Marcet, *Op. cit.* (45), pp. 7–10

56. Rayner-Canham, *Op. cit.* (3) p. 16

57. H. J. Abrahams, "The chemical library of Thomas Jefferson", *J. Chem. Educ.*, 37, 1960, 357–360

58. Quoted in reference 41

59. Anne Treneer, *The Mercurial Chemist – A life of Sir Humphry Davy*, Methuen, London, 1963, p. 135

60. A. de la Rive, "Madame Marcet", *Bibliothèque Universelle – Revue Suisse et Étrangère*, 1859, 64(4), 445–468

61. N. Pigeard, "Chemistry for Women in Nineteenth-Century France", *in* Anders Lundgren, Bernadette Bensaude-Vincent (Eds.), *Communicating Chemistry – Textbooks and Their Audiences, 1789–1939*, Science History Publications/ Canton, MA, 2000, p. 313

62. Jane Marcet, *Conversations on Chemistry*, vol. 2, 5th edition, Longman, London, 1817, p. 362

63. Marcet, *Op. cit.* (45), pp. 19–20

64. Louis Aimé-Martin, *Lettres à Sophie sur la Physique, la Chimie et l'Histoire Naturelle*, Gosselin, Paris, 1822 (2nd edition), vol. 1, pp. 81–84

65. Pigeard, *Op. cit.* (61), p. 315

66. The complete title is *A Chimica Ensinada em 26 Liçoens: Contendo o desenvolvimento, e theorias desta sciencia, postas ao alcance de toda a gente, e a cada lição correspondem experiências chimicas e applicaçoens ás artes.*

67. *La Chimie enseignée en 26 leçons: contenant le développement des théories de cette science, mises à la portée des gens du monde, et a chaque leçon des expériences chimiques et des applications aux arts*

68. M. Payen, *La Chimie enseignée en 26 leçons*, P. J. de Mat, Bruxelles, 1825, p. 201–203

69. I. M. Peres, S. P. J. Rodrigues, "De Jane Marcet ao Visconde de Vilarinho de São Romão: conversas sobre química no século XIX", *História, Ciências, Saúde – Manguinhos*, 25 (2018) 469–495

70. Rayner-Canham, *Op. cit.* (3), pp. 18–19

71. G. Jeffery Leigh, A. J. Rocke, "Women and Chemistry in Regency England: New Light on the Marcet Circle", *Ambix*, 63 (2016) 28–45

72. Quoted in reference 71

73. H. S. van Klooster, "Amos Eaton as a Chemist", *J. Chem. Educ.*, 15, 1938, 453–460

74. D. J. Warner, "Science Education for Women in Antebellum America", *Isis*, 69 (1978) 58–67

75. M. E. Weeks, F. B. Dains, "Mrs. A. H. Lincoln Phelps and her services to chemical education", *J. Chem. Educ.*, 14 (1937) 53–57

76. S. Badilescu, "Chemistry for beginners. Women authors and illustrators of early chemistry textbooks", *Chem. Educator*, 6, (2001), 114–120

77. Almira H. Lincoln Phelps, *Chemistry for Beginners*, F. J. Huntington, New York., 1842, p. 3

78. *Ibid.*, p. 10

79. *Ibid.*, n.p.

80. Almira H. Lincoln Phelps, *Chemistry for Beginners*, F. J. Huntington, Harford, 1834, p. 119

81. F. D. White, "«Sweet Skepticism of the Heart»: Science in the Poetry of Emily Dickinson", *College Literature*, 19 (1992), 121–128

82. Richard B. Sewall, *The life of Emily Dickinson*, Harvard University Press, Cambridge (MA), 2003, pp. 362, 367

83. H. Uno, "«Chemical Conviction»: Dickinson, Hitchcock and the Poetry of Science", *The Emily Dickinson Journal*, 7 (1998) 95–111

84. Benjamin Silliman, Jr, *First principles of chemistry for the use of colleges and schools*, H. C. Peck & Theo. Bliss., Philadelphia, 1854, 33ª edição, pp. 192–198

85. Phelps, *Op. cit.* (77), p. 135

86. Ihde, *Op. cit.* (37), p. 191

87. E. C. Patterson, "The Case of Mary Somerville: An Aspect of Nineteenth-Century Science", *Proc. Am. Philos. Soc.*, 118 (1974) 269–275

88. Mary Somerville, *On molecular and microscopic science*, vol. I, John Murray, London, 1869

89. *Ibid.*, p. 100

90. E. C. Patterson, "Mary Somerville", *British Journal for the History of Science*, 4 (1969) 311–339

91. Quoted in reference 87

92. B. T. Gates, "Revisioning Darwin with sympathy – Arabella Buckley", em *Natural Eloquence – Women reinscribe science*, B. T. Gates, A. B. Shtein (eds), The University of Wisconsin Press, Madison, 1997, pp. 164–176

93. Melane Keen, *Science in Wonder Land – The scientific fairy tales of Victorian Britain*, Oxford University Press, Oxford, 2015, pp. 99–100

94. R. Somerset, "Arabella Buckley and the Feminization of Evolution as a Communication Strategy", *Nineteenth-century Gender Studies*, 7 (2011), n.p.

95. B. Lightman, "The Story of Nature: Victorian Popularizers and Scientific Narrative", *Victorian Review*, 25 (2000), pp. 1–29

96. S. Reisert, "The Magic of It All: How Victorians found a foolproof way to make science interesting for their children", *Distillations*, 2 (2016) 44–45

97. Lily Martin, *Princess & Fairy, or, The Wonders of Nature*, W. & R. Chambers, London, 1899, p. 57

98. M. R. S. Creese, "Martha Annie Whiteley (1866–1956): chemist and editor", *Bull. Hist. Chem.*, 20 (1997) 42–45

99. Marelene F. Rayner-Canham, *Women in Chemistry: Their Changing Roles from Aalchemical Times to the Mid-twentieth Century*, Chemical Heritage Foundation, Philadelphia, 2001, pp. 192–193

100. S. Pramer, "Mary Fieser – A transitional figure in the history of women", *Journal of Chemical Education*, 62 (1985) 186–191

101. Susan Ware (ed.), *Notable American Women: A Biographical Dictionary Completing the Twentieth Century*, Belknap Press of Harvard University Press, Cambridge (MA), 2004, pp. 207–208

102. Rayner-Canham, *Op. cit.* (99), p. 194

103. *"Chemistry Department creates Fieser Fellowship"*, *Harvard Gazette,* February 7, 2008, https://news.harvard.edu/gazette/story/2008/02/chemistry-department-creates-fieser-fellowship/

104. L. F. Fieser, "Synthesis of Vitamin K_1", *Journal of American Chemical Society*, 61 (1939) 3467–3475

Index

Printed in the United States
by Baker & Taylor Publisher Services